GLOBAL DIGITAL TRANSFORMATION AND THE COVID-19 PANDEMIC

GLOBAL DIGITAL TRANSFORMATION AND THE COVID-19 PANDEMIC

Edited by

Rajdeep Chowdhury, PhD

K. Martin Sagayam, PhD

Neyara Radwan, PhD

Ahmed A. Elngar, PhD

APPLE ACADEMIC PRESS

First edition published 2025

Apple Academic Press Inc.
1265 Goldenrod Circle, NE,
Palm Bay, FL 32905 USA

760 Laurentian Drive, Unit 19,
Burlington, ON L7N 0A4, CANADA

CRC Press
2385 NW Executive Center Drive,
Suite 320, Boca Raton FL 33431

4 Park Square, Milton Park,
Abingdon, Oxon, OX14 4RN UK

Library and Archives Canada Cataloguing in Publication

Title: Global digital transformation and the COVID-19 pandemic / edited by Rajdeep Chowdhury, K. Martin Sagayam, Neyara Radwan, Ahmed A. Elngar.

Names: Chowdhury, Rajdeep, editor. | Sagayam, K. Martin, 1987- editor. | Radwan, Neyara, 1975- editor. | Elngar, Ahmed A., editor.

Description: First edition. | Includes bibliographical references and index.

Identifiers: Canadiana (print) 20240400348 | Canadiana (ebook) 20240400380 | ISBN 9781774915509 (hardcover) | ISBN 9781774915516 (softcover) | ISBN 9781032716862 (ebook)

Subjects: LCSH: Medical technology. | LCSH: Artificial intelligence—Medical applications. | LCSH: Machine learning. | LCSH: Internet of things. | LCSH: COVID-19 Pandemic, 2020-

Classification: LCC R855.3 .G56 2025 | DDC 610.285—dc23

Library of Congress Cataloging-in-Publication Data

CIP data on file with US Library of Congress

ISBN: 978-1-77491-550-9 (hbk)
ISBN: 978-1-77491-551-6 (pbk)
ISBN: 978-1-03271-686-2 (ebk)

About the Editors

Rajdeep Chowdhury, PhD
Professor, Computer Application; Life Member, Computer Society of India, India (1982–2020)

Rajdeep Chowdhury, PhD (1982–2020), was an eminent academician, trend-setting author, and award-winning poet. He had more than 12 years working experience in academics. His research areas included cryptography, data warehouse, data mining, network security, etc. He authored and presented more than 90 research papers in international and national journals and conferences, He has authored/edited 15 books with seven others in preparation. He has also been program committee member for more than 120 international conferences in various parts of the world and advisory board member for numerous international conferences. Dr. Chowdhury was on the editorial boards of more than 30 international journals and holds three published patents. His articles, poems, and stories have been widely anthologized in more than 60 anthologies, e–zines and periodicals, both in India and abroad. Dr. Chowdhury's authored innovative and unique book POISINE: Bengali Cuisine with Poetry (Level 1: Starters) is in Asia Pacific Records. He delivered more than 30 keynote addresses and invited lectures at international and national conferences, seminars, and symposiums all over the globe. Dr. Chowdhury was the recipient of many prestigious awards and recognitions.

K. Martin Sagayam, PhD
Assistant Professor, Department of ECE, Karunya Institute of Technology & Sciences (Deemed to be University), India

K. Martin Sagayam, PhD, is currently working as Assistant Professor in the Department of ECE, Karunya Institute of Technology & Sciences (Deemed to be University), India. He has authored/coauthored a number of papers in refereed international journals. He has also presented many papers at reputed international and national conferences. He has authored three edited books, three authored books, book series, and more than 15 book chapters with reputed international publishers such as Elsevier, Springer, IGI Global, and CRC Press. He has three Indian patents and three Australian patents for his innovations. He has been appreciated as the best reviewer for the *Journal of Signal, Image, and Video Processing, Intelligent Decision Technologies, International Journal of Engineering Research & Technology, Pattern Analysis and Its Applications,* and *Informing Science: The International Journal of an Emerging Transdiscipline*. He is an active member of professional bodies such as IEEE, Engineering and Scientific Research Groups, International Society of Promising Computer Engineers, Copernicus, Scientific Engineering Research Corporation, International Association of Computer Science and Information Technology, International Association of Engineers, Indian Society of Electronics and Communication Engineering, and ORCID. His area of interest includes communication systems, signal and image processing, machine learning, and virtual reality. His PhD is in the field of image processing, pattern recognition, and machine learning.

Neyara Radwan, PhD
Associate Professor, Mechanical Engineering, Faculty of Engineering, Suez Canal University, Egypt & Associate Professor & Quality Assurance Officer of the PhD Program King Abdulaziz University, Saudi Arabia

Neyara Radwan, PhD, is a staff member in two universities in two different countries: Egypt and Saudi Arabia. She is working as Associate Professor and Quality Assurance Officer for the PhD program at King Abdulaziz University in Saudi Arabia, as well as in the Mechanical Department, Faculty of Engineering, Suez Canal University, Egypt. She has participated in more than 40 international conferences as a keynote speaker in many countries such as Singapore, Italy, India, the United States, the Philippines, and more. She has collaborated on many academic international projects and published many papers in high-quality peer-reviewed journals. Moreover, she serves as an editorial board member as well as a reviewer for many high-quality international academic journals. She is an active member in many international academic communities such as a member of the Board of Directors of Accord University, Somalia, a member of Asia Logistics and Supply Chain Council (ALSCC), Malaysia. She is also a member of the Global

Education Policy Network (GEPN), which is sponsored by Prince Sultan University (Saudi Arabia) and The State University of New York at Buffalo (the United States), 2020. She has been selected to be a member of the World Association of Professional Scientistic (WAPS), Springer. In addition, she is a country advisor in Positive Thoughts, a nonprofit organization in India. She has a full membership in the Organization for Women in Science for the Developing World (OWSD), based in Italy. In 2022, she was nominated from G100 as the Country Chair for Saudi Arabia for the Engineering Wing. In 2022, she was chosen to be the Egyptian Ambassador in the International Human Rights Commission Organization based in Denmark

Ahmed A. Elngar, PhD
Associate Professor of Computer Science
Faculty of Computers and Artificial Intelligence,
Beni-Suef University, Beni-Suef City, Egypt
College of Computer Information Technology,
American University in the Emirates, United Arab Emirates
Chair of Scientific Innovation Research Group (SIRG)

Ahmed A. Elngar, PhD, is the Founder and Head of Scientific Innovation Research Group (SIRG). He is Associate Professor of Computer Science and Artificial Intelligence, Beni-Suef University, Beni-Suef City, Egypt. He is also Associate Professor of Computer Science at College of Computer Information Technology, American University in the Emirates, United Arab Emirates. Dr. Elngar is a Director of the Technological and Informatics Studies Center (TISC), Faculty of Computers and Information, Beni-Suef University. He has more than 60 scientific research papers published in prestigious international journals and over 25 books covering such diverse topics as data mining, intelligent systems, social networks, and smart environment. Dr. Elngar is a collaborative researcher and a member of the Egyptian Mathematical Society (EMS) and International Rough Set Society (IRSS). His research areas include Internet of Things (IoT), network security, intrusion detection, machine learning, data mining, artificial intelligence, big data, authentication, cryptology, healthcare systems, and automation systems. He is an editor and reviewer of many international journals around the world. Dr. Elngar has won several awards including the Young Researcher in Computer

Science Engineering, from Global Outreach Education Summit and Awards 2019, on January 31, 2019 at Delhi, India. He was also awarded the Best Young Researcher Award (male) (below 40 years), at the Global Education and Corporate Leadership Awards (GECL-2018), Rajasthan, India. He also has an low called "ElDahshan Authentication Protocol," Information Technology Industry Development Agency (ITIDA), Technical Report, 2016. Dr. Elngar has performed many great activities for the benefit of the community and the environment including organizing 12 workshops hosted by a large number of universities in almost all governorates of Egypt.

Contents

Contributors

Alankrita Aggarwal
Panipat Institute of Engineering and Technology, Samalkha, Haryana, India

Noor Al Ali
American University of the Emirates, Dubai, UAE

Nitin Arora
Electronics & Computer Discipline, Indian Institute of Technology, Roorkee, India

Andi Asrifan
Universitas Muhammadiyah Sidenreng Rappang, Indonesia

Boudhayan Bandyopadhyay
Department of Biotechnology, School of Life Science and Biotechnology, Adamas University Kolkata, India

Ajit Behera
Department of Metallurgical & Materials Engineering, National Institute of Technology, Rourkela, Odisha, India

Ankan Bhaumika
Mathematics Department, Dhamcha Chhagulia Siddheswari High School, Paschim Medinipur, West Bengal, India

Muhammad Bilal
School of Life Science and Food Engineering, Huaiyin Institute of Technology, Huaian, China

Suman Chatterjee
Department of Mechanical Engineering, National Institute of Technology, Rourkela, Odisha, India

Lenny Christopher
Department of Bioinformatics, Karunya Institute of Technology and Sciences, Coimbatore, Tamil Nadu, India

Pradeep K. Das
Department of Chemistry, N. C. (Autonomous) College, Jaipur, Odisha, India

Shankar Debnath
Computer Science and Engineering Department, JIS College of Engineering, Kalyani, India

Kuldeep Dhama
Division of Pathology, ICAR-Indian Veterinary Research Institute, Izatnagar, Bareilly, Uttar Pradesh, India

Ahmed A. Elngar
Faculty of Computer & Artificial Intelligence, Beni-Suef University, Beni-Suef City, Egypt
College of Computer Information Technology, American University in the Emirates, United Arab Emirates

R. Jennifer Eunice
Department of Electronics and Communication Engineering, Karunya Institute of Technology and Sciences, Coimbatore, Tamil Nadu, India

Shivani Gaba
Panipat Institute of Engineering and Technology, Samalkha, Haryana, India

Lakshay Goyal
School of Computer Science, University of Petroleum & Energy Studies, Dehradun, India

Poonam Gupta
G H Raisoni College of Engineering and Management, Pune, Maharashtra, India

Om Prakash Jena
Department of Computer Science, Ravenshaw University, Odisha, India

Aditya Kakde
School of Computer Science, University of Petroleum & Energy Studies, Dehradun, India

Sharun Khan
Division of Surgery, ICAR-Indian Veterinary Research Institute, Izatnagar, Bareilly, Uttar Pradesh, India

Salahddine Krit
Faculty of Sciences, University Ibn Zohr, Agadir, Morocco

Indhraom Prabha M.
G H Raisoni College of Engineering and Management, Pune, Maharashtra, India

Manas Kumar Mishra
School of Computer Science and Engineering, VIT Bhopal University, Kothri Kalan, Madhya Pradesh, India

Ranjan K. Mohapatra
Department of Chemistry, Government College of Engineering, Keonjhar, Odisha, India

Khalid Atiq Hassan Mubarak
American University of the Emirates, Dubai, UAE

Muthmainnah
Universitas Al Asyariah Mandar, Indonesia

Huma Naaz
School of Computer Science, University of Petroleum & Energy Studies, Dehradun, India

J. Andrew Onesimu
Department of Computer Science and Engineering, Manipal Institute of Technology, Manipal Academy of Higher Education, Manipal, Karnataka, India

Madhumita Pal
Department of Electrical Engineering, Government College of Engineering, Keonjhar, Odisha, India

Paryati
ST., M.KOM (Doctoral Candidate in UITM Malaysia), Lecture University Pembangunan National "Veteran" Yogyakarta, Indonesia

Garima Pandey
Department of Computer Science, National Institute of Technology Karnataka Surathkal, India

Claudiu-Ionut Popirlan
Department of Computer Science, Faculty of Sciences, University of Craiova, Romania

Cristina Popirlan
Department of Computer Science, Faculty of Sciences, University of Craiova, Romania

Shriram Raghunathan
School of Computer Science and Engineering, VIT Bhopal University, Kothri Kalan, Madhya Pradesh, India

Pranati Rakshit
Computer Science and Engineering Department, JIS College of Engineering, Kalyani, India

Karthikeyan S.
Centre for Innovation and Entrepreneurship, Krishnasamy College of Engineering and Technology, Cuddalore, India

Pronit Sarkar
Computer Science and Engineering Department, JIS College of Engineering, Kalyani, India

Adriana Burlea Schiopoiu
Department of Management, Marketing, Business Administration, Faculty of Economy and Business Administration, University of Craiova, Romania

Edyta Skibinska
American University of the Emirates, Dubai, UAE

Shashwat Tiwari
School of Computer Science, University of Petroleum & Energy Studies, Dehradun, India

Baranidharan V.
Department of Electronics and Communication Engineering, Bannari Amman Institute of Technology, Sathy, India

Ahmad Al Yakin
Universitas Al Asyariah Mandar, Indonesia

Abbreviations

AI	artificial intelligence
ANN	artificial neural network
CDC	Centre for Disease Control
COVID-19	Corona virus 2019
CoV-HKU1	Coronavirus HKU1
CSE	computer science engineering
CSSE	Center for Systems Science and Engineering
CT	computed tomography
DHA	Dubai Health Authority
DHCC	Dubai Healthcare City
ECG	electrocardiogram
EDA	electrodermal activity
EEG	electroencephalogram
EHRs	electronic health records
EMG	electromyography
ESL	English as a second language
FIPV	Feline Infectious Peritonitis Virus
GIS	geographic information system
GNSS	global navigation satellite systems
GoI	Government of India
GSM	global system for mobile communication
HCI	human–computer interaction
HCoV-229	Human Corona Virus 229
HCoV-NL63	Human Corona Virus NL 63
HDS	hybrid diagnose strategy
ICDs	implantable cardioverter-defibrillators
ICMR	Indian Council of Medical Research
ICT	information and communication technology
IDF	International Diabetes Foundation
IDS	intrusion detection system
IFS	intuitionistic fuzzy set
IHR	International Health Regulations
IoMT	internet of things for medical things
IT	information technology
IVSFS	interval-valued spherical fuzzy set
IVRS	interactive voice response system
JADE	Java Agent Development

LoRa	low range
LSTM	long–short-term memory
MBKM	Merdeka Belajar Kampus Merdeka
MCU	monitoring control unit
MDCNN	modified deep convolution neural network
MeitY	Ministry of Electronics and Information Technology
MERS	Middle East respiratory syndrome
MoHaP	Ministry of Health and Prevention
MoHFM	Ministry of Health, Family, and Welfare
MSE	mean square error
NBTC	National Broadcasting and Telecommunication
nCOV	novel corona
NHS	National Health Service
NIC	National Informatics Centre
NIV	National Institute of virology
PEDV	porcine epidemic diarrhea virus
PHEIC	Public Health Emergency of International Concern
RF	radio frequency
ROI	region of interest
RT-PCR	reverse transcription-polymerase chain reaction
RVNS	reduced variable neighborhood search
SARS	severe acute respiratory syndrome
SARS-CoV-1	severe acute respiratory syndrome coronavirus
SDLC	Software Development Life Cycle
SFS	spherical fuzzy set
SLM	severity level measure
SMT	surface mount technologies
UGC	user-generated content
USB	Universal Serial Bus
VoIP	voice over Internet Protocol
WHO	World Health Organization

Preface

Information science is a discipline primarily concerned with the analysis, collection, classification, manipulation, retrieval, navigation, and dissemination of information. Information science focuses on understanding problems from the perspective of relevant stakeholders and applying information and other technologies where necessary. This means addressing system issues first rather than individual technology components within the system.

The focus of this book is on the adoption of artificial intelligence (AI), machine learning (ML), and Internet of Things (IoT) technologies that are effective in the medical context. It is used to improve treatment efficiency, avoid the risk of false diagnosis, make therapeutic decisions, and predict the outcome of many clinical scenarios, especially the COVID-19 pandemic.

The book begins with the fundamentals of computer intelligence, related techniques, and procedures. This book contains cutting-edge methods of new information technology used in healthcare, addressing the problem of effective data analysis and storage in healthcare institutions.

Therefore, this book provides an excellent platform for sharing knowledge among academics, engineers, industry professionals, researchers, scientists, and students working in the fields of data and information science and provides a description of their work. It aims to provide interdisciplinary and transdisciplinary research advances related to mining. The book's chapter selection will inspire aspiring researchers, especially those interested in and involved in interdisciplinary research related to COVID-19. The main focus of the proposed book is to create a sound foundation for exchanging knowledge and skills, ideas, the latest trends, and insights of researchers.

An overview of the topics addressed in each chapter is given below:

- Chapter 1 describes the technical practices of different countries and integrates them by embedding biomicrochips in quarantine in the early stages using the Global Systems for Mobile Communications (GSM) and Global Navigation Satellite Systems (GNSS) modules. The recommended approach includes these for effective monitoring and spread control. Using this approach will bring significant benefits to health, authorities, and government authorities in reducing the risk of rapid spread of disease. It aims to ensure immediate follow-up with more efficient tracking, along with an appropriate warning mechanism.

- The purpose of Chapter 2 is to use a learning management system to develop learning design models based on mobile learning and to develop mobile services, which are web-based mobile service applications. The phase of creating an application focuses on developing a mobile learning model. To do this, we must create a mobile learning application software that already contains some learning material content, and then develop a design model for the learning application. The application is developed on mobile servers with web-based applications for mobile services.
- Chapter 3 provides details on the outbreak of COVID-19, the facts of the world, and more specifically India. We have introduced a detailed case study of human-to-human transmission of COVID-19. This chapter also includes some analysis, including: status-related infections, infection rate charts, COVID-19 case fatality charts, etc. In addition, a deep learning model was used to experiment with COVID-19 chest X-rays to detect COVID-19 infection. Based on our experimental results, we can conclude that the model performs remarkably in detecting COVID-19 infection. Nevertheless, attempts are being made to improve the model and test new forms of research to improve the criteria. The situation remains the same until we find new ways to control this deadly coronavirus. Immunization is a viable option as it boosts immunity and should be administered to control this pandemic.
- Chapter 4 describes the COVID-19 transport problem between interval-valued language spherical fuzzy sets. The COVID-19 shipping issue is resolved in this chapter using the modified Warshall–Floyd algorithm. This chapter uses a new asphericization approach. Approaches to solving the real problem show a flaw in government decisions related to public transport in the COVID-19 situation. The main advantage of this proposed method is the simple calculation and the assumption of the language form of the variable to be considered, but the main drawback of this proposed method is the elaborate time-consumption during the calculation. Finally, there is an urgent need for surveys on attitudes and behaviors and further analysis of management decisions for transportation systems during the pandemic, and this study will help identify the shortest route

between different locations to minimize the spread of the virus. It helps you design an effective public transportation system to use.

- Chapter 5 presents an intelligent agent system implemented to manage the nursing process in a real environment. The intelligent agent itself will be developed using an open source platform for agent-based peer-to-peer software applications called the JADE-JAVA agent development framework. It combines a suite of intelligent agent technologies. Incomplete and contradictory information, the concept of negotiation and accountability for the distribution of tasks, the concept of obligations and conventions for managing coherent collaboration, and a set of communication primitives for interaction between agents are discussed. The utility of this approach will be demonstrated through the development of an application prototype for the clinical management process of COVID-19 treatment.
- In Chapter 6, the network model proposed is rational and suitable for predicting cases of COVID-19. Accurately predicting illness helps provide quality medical services. Therefore, we hope that this study can provide useful information for monitoring the outbreak of COVID-19.
- Chapter 7 covers various applications and implementation models, new fog and edge computing beyond traditional cloud models, key challenges facing IoT in healthcare, and IoT platform security practices and issues. This chapter also covers common security attacks that take place in IoT environments. Additionally, this chapter reveals how IoT has risen to deal with the current pandemic situation in this fight against the deadly COVID-19 virus. Finally, the future growth lines of IoT including enabling technologies are discussed.
- Chapter 8 presents studies that used the K-nearest neighbor regressor and the gradient boosting regressor to create predictive models for confirmed cases, recoveries, deaths, or mortality. The model performance is very good at predicting all cases. The R-squared value is very close to 1. Future directions of this work will analyze the steps taken by different countries in their policy measures. This kind of analysis can help other countries change their policies to combat this kind of pandemic.
- In Chapter 9, it is argued that Mucor disease, known as black fungus, is a rare and invasive fungal infection that often targets

immunocompromised patients. A group of molds called Mucors causes mucordosis. According to the US Centers for Disease Control and Prevention (CDC), the mortality rate for this type of infection is 46%. According to a survey, only 109,000 people in India have suffered from zygomycosis each year in the past, which is equivalent to 1% of the total population. However, recently, the number of cases is increasing at a very fast pace. Uncontrolled diabetes and inappropriate steroid use are believed to be two major factors in the prevalence of mucormycosis in India. The fungus spreads through the sinuses to the eye sockets and affects the eyes, brain, and lungs. Therefore, it is important to diagnose mucormycosis at the right time for timely treatment. Therefore, this chapter provides insight into the occurrence of mucormycosis in diabetic and COVID-19 patients, describes the type of disease (mucormycosis/COVID-19), and explains how these molds enter the body of immunocompromised patients. In addition, this chapter discusses possible treatments and preventive measures.

10. Chapter 10 describes how cybersecurity is becoming a major concern in the digital world. This is explained in medical care by the value that medical records hold containing a wealth of personal information (information that may put patients at risk of data breach). Healthcare providers consider who has access to that knowledge and consider the administrator of patient information when it is necessary to secure the patient's medical records. In line with these facts, UAE will not only protect patient data by law, but also provide a unified system of records through the Riayati project to reduce the risk of malpractice and make the healthcare system more effective. The vision has started. By providing accurate and up-to-date patient information to healthcare providers. This study presents a case study of the Riayati project conducted in the United Arab Emirates to improve the quality, safety, and cost of healthcare. Interviews, readings, and observations have been used to collect the data, allowing researchers to delve into the context of the topic to observe reality and gather information. The authors used this research method because the topic is understudied and serves as a starting point for future research.

11. Chapter 11 describes standard teaching and learning methods that do not improve students' cognitive and communicative abilities.

Modern ICT tools can be used to improve the efficiency of teaching and learning processes and educational systems. A literature review reveals that the integration of current ICT tools into the teaching and learning process significantly improves the teaching and learning process. The COVID-19 epidemic has changed people's lives in all areas, including schooling. To comply with health protocols, the learning process cannot be done in person. Instead, it is conducted primarily online. This study aims to examine contemporary information and communication technology tools and their use in the teaching and learning process. Consequently, this study investigated how learning processes are implemented in online learning-based educational courses. The study sample consisted of 42 randomly selected respondents from 12 courses. The data collection tool used in this study was an online questionnaire and the results were analyzed descriptively. The results show that course teachers frequently use online communication tools, with WhaYoZo being the most used, followed by media and technology teaching tools. The results of this study show that higher education needs to support and encourage SOCMED-WhaYoZo to improve students' technical skills, performance, expression of ideas, reasoning, and achievement for sustainable development.

- In Chapter 12, students expressed that self-motivation, goal orientation, content and technology are essential and important aspects to determine the success of online training platforms during COVID-19 lockdowns and this is explained here. Our research demonstrates the problems and challenges faced by students arriving for internships during COVID-19 lockdowns. We believe our research will provide important insights to both academia and the student community regarding the challenges students face during internships, and provide useful strategies to increase their self-motivation.

- Chapter 13 states that in the current pandemic situation, even after thorough investigation and hard work, we have not found a permanent end to this COVID-19 virus and the safety of all is at stake. Besides, you can't keep everyone at home. Most tasks have moved home from offices, but the economy shows the devastating effects of the current situation. A safe information system is proposed in this chapter to help people avoid high risk areas. Post-pandemic crowd safety is about safely informing users about the current environment by providing detailed information about the current pandemic situation. Thus, users can finally understand how many coronavirus hotspots exist and in which areas they are predominantly located, and be sure to avoid them. This application has saved many lives and even helps reduce the spread of the coronavirus and other similar viruses by warning people in advance.

Finally, we would like to express our sincere gratitude to the authors of all the chapters for providing time and expertise and contributing to the success of the book. Editors also want to recognize the valuable contributions of reviewers in improving the quality, consistency, and presentation of content in chapters.

CHAPTER 1

Use of Communication Technology for Combating the COVID-19 Pandemic

BARANIDHARAN V.[1] and KARTHIKEYAN S.[2]

[1]*Department of Electronics and Communication Engineering, Bannari Amman Institute of Technology, Sathy, India*

[2]*Centre for Innovation and Entrepreneurship, Krishnasamy College of Engineering and Technology, Cuddalore, India*

ABSTRACT

The outbreak of the COVID-19 virus has been rapid in many countries with no vaccine and specific medicine for treatment which is under research. Thus, self-quarantine and social distancing are perceived to be highly effective in breaking the chain of onward spread. Despite adverse effects and serious efforts, lack of cooperation and a few quarantines without realizing the consequences are consistently reported across the globe. As recommended by WHO, tracking, testing, and treating COVID-19-infected patients is the globally recommended practice, many countries have utilized the development of communication technologies such as mobile communications, Bluetooth, global positioning systems, and facial recognition, timestamps, and wristbands for tracking and contact tracing. All these technologies require the active participation of quarantined people. In this chapter, we have discussed the technological practices of various countries and recommended an integrated approach by involving the global system for mobile communication (GSM) and global navigational satellite system (GNSS) module along with implanting of Bio-Microchips in the quarantine during the early stages for effective monitoring and control of the spread. Immediate tracking with more efficient tracing along with a proper alert mechanism shall be ensured

Global Digital Transformation and the Covid-19 Pandemic.
Rajdeep Chowdhury, K. Martin Sagayam, Neyara Radwan, Ahmad A. Elngar (Eds.)

by making use of this approach which will be highly useful to the health officials, authorities, and administrators in reducing the risk of rapid spreading.

1.1 INTRODUCTION

The World Health Organization (WHO) stated the outbreak of Novel Corona (nCOV) virus 2019 (COVID-19) as a Public Health Emergency of International Concern (PHEIC) on January 30, 2020, based on the emergency committee advice under International Health Regulations (IHR).[1] It is evident that the spreading of COVID-19 is majorly related to international traffic, especially in countries with infection history. It is imperative that the control measures at the early containment phase of an outbreak are based on the public health rationale.

Most of the affected countries are attempting various self-sustained measures for controlling the rate of spread. The rest of the countries are expected to be prepared with effective measures. Based on the evolving situation, the restrictions are required to be in proportionate to their public health risk. WHO advised all the countries to be prepared for containment, active surveillance, early detection of COVID-19, isolation of nCOV-infected persons, tracing of their contacts, and prevention of the onward spread of infection. The transmission of the nCOV virus is significantly associated with international traffic except in the country of origin.[2]

Many countries have decided to repatriate their nationals from Wuhan City and other affected countries.[3] Some of the key considerations are required to be made, in order to ensure the health and wellbeing of those involved in the repatriation. The WHO continuously advises trade restrictions to the countries experiencing the COVID-19 outbreaks. The evidence shows that restrictions to the people's movements during such public health emergency are practiced inefficiently due to the lack of public cooperation in many situations. In general, the governments of all the countries are finding it difficult to impose certain restrictions on socio-economic and political grounds. However, such restrictions shall be enforced as required by the affected countries with the aid of technology to avoid the rapid onward spreading of COVID-19.[4]

The travel bans from the affected countries or passenger entry denial to prevent the importation of the virus from the infected cases is not an easy solution as it may have a significant economic and social impact.[5] Temperature screening at entry or exit places in airports/railway stations/ports alone is not an effective way to stop COVID-19 from spreading. Data collection including health declarations and contact details of travelers at arrival has become a most common practice. The travelers from the COVID-19 affected countries must be quarantined for up to 14 days as per the guidance.[6] In spite of all efforts, some quarantined people are moving freely from one place to another without realizing the consequences of COVID-19 spreading and outbreak.

As the entire world attempts to combat the COVID-19 pandemic, many countries are making serious efforts to track the people with a history of travel. In combating the spread of COVID-19, different technologies such as Android App, Robotic bots for treating the patients, time-stamped photos-based tags, facial recognition techniques, and Network stations are used. Many government bodies are attempting to use the mobile apps and the other technologies stated above in order to track the persons with international travel history. It is obvious that the latest breakthrough in communication technologies has enabled rapid contact tracing for effective control. Apart from the technological shifts need to be employed to track the quarantined traveled peoples.

This chapter is organized as follows: The next section reviews communication technologies used by the different countries to control the spreading of COVID-19 and the subsequent section discusses an enhanced approach for effective tracking with concluding remarks in the end.

1.2 TECHNOLOGICAL PRACTICES ADOPTED WORLDWIDE

Technology aided tracking, quarantine status monitoring, and contact tracing practices of a few countries in view of preventing rapid onward spreading of COVID-19 have been discussed below.

1.2.1 CHINA

In Ruichang city of China, the police and security officers are using satellite system-enabled copters or drones for monitoring the high dense population and congested areas. The Zhuhai-I Hyper Spectral Imaging Satellites systems resulted in effective non-stop monitoring of remote areas. TESTER, a second-generation artificial intelligence (AI)-based satellite designed by the Satellite Technology and ADA space, is capable of powerful analyzing and processing of data to predict the spreading of the virus. By combining all the informatics, the people could visualize the geographical outreach of COVID-19 graphically.[7]

The Chinese government in collaboration with tech giants developed a color code-based health rating mobile App that tracks millions of people for continuous and vigorous monitoring. This smartphone App assigns three different colors on the basis of travel, medical histories, and the persons to be allowed in public spaces was decided based on the color codes. Several organizations developed many dashboards by using the concept of big data analytics to access public information to enable continuous monitoring. Some of the leading Chinese AI companies have come up with special computer vision technology that can even monitor the wearing of masks.

1.2.2 GERMANY

The health authorities in Germany turned to a new fitness tracking app named "Corona datenspende" that translates and gather the nationwide data for intensive monitoring of COVID-19 spreading.[8] The user information collected by the App includes automatic and manually recorded activities like walking, blood pressure, heart rate, and socio-demographic data. Further, the German Government is urging the use of AI-based applications to find the rate of spreading of COVID-19 and also for rendering advice during the home quarantine by using voice assistance.

1.2.3 HONG KONG

The administration of Hong Kong has taken the initiative of strapping the electronic wristbands on all the arriving passengers from the COVID-19 affected countries. This technique ensures that they have complied with the mandatory 14 days of quarantine. Such people are instructed to walk along with the corners of their house so that the App can track the coordinates of the living space where they will be quarantined.[9] Each wristband will have a QR code linked to the App to ensure the wearer's safe stay within the home. However, the wristband does not track the location; instead, it identifies only the changes in location. The attractive feature is that whenever the wristband is broken or the smartphone is disconnected or removed from the living space, this App will send alert messages to the respective authorities.

1.2.4 INDIA

The Ministry of Electronics and Information Technology (MeitY) and the Government National Informatics Centre (NIC) of the Government of India have launched a mobile app named "Aarogya Setu" for connecting the essential health services with the people of India to fight against COVID-19.[10] This mobile app is aimed at augmenting the initiatives of the Department of Health, Government of India to proactively reach out and inform the users. This App is used to know about risks, best practices, and relevant advisories pertaining to the containment of COVID-19. This App collects the user's mobile number, and location data (via GPS and Bluetooth) and matches their travel history with that of the COVID-19 positive patients. Such data are documented by the Indian Council of Medical Research (ICMR) in the backend.

With the help of GPS and Bluetooth, location-based social demographics are generated and advise the user not to venture out into a containment zone. Further, this App allows the users to check whether they have accidentally or unknowingly encountered any infected persons. This App is also used as a clarifying tool by asking the users to answer a few questions related to their symptoms pertaining to COVID-19, and the information will be stored on the Government server for further processing. Furthermore, this App uses the Bluetooth signal between the phones like Singapore's trace together app which helps the authorities in contact tracing.

The Government of Tamil Nadu state of India has launched an interactive voice response system (IVRS) for the citizens.[11] This facility is handy, especially for those who are not using smartphones. A toll-free number to be called by the citizens is advertised and the system responds back by asking simple questions connected to the symptoms of the COVID-19 pandemic. The caller has to simply press the respective digits against the options given depending upon their symptoms. Based on the answers, the callers have been categorized in the "No Risk," "Low Risk" and "High Risk" categories. Accordingly, the district team will reach out to them, to provide necessary support in the form of medication, quarantine, or hospitalization as required. The main focus of this IVRS facility is to provide support to the Department of Health in the form of providing data and geo-fencing spatial datasets for further decisions. In India, a panel of Joint secretaries at the Ministry of Home Affairs has proposed to use technologies such as big-data analytics, AI, and geo-fencing techniques in more effective and efficient contact tracing.[12]

1.2.5 ISRAEL

The Israel Government has approved Israel Security Agency on March 17, 2020, to monitor suspected human virus carriers. On March 22, 2020, the government has launched a new App called "Hamagen," this app allows all the users to know their status of crossing paths with infected persons.[13] This App can be installed either on iOS or any other Android device. This App also reports the movements of the quarantined people. Such information stored for retrieval and sent to respective authorities unilaterally for the purpose of cross-reference.

1.2.6 POLAND

The Polish Ministry of Digital Affairs, the Republic of Poland, launched a home quarantine application on March 2020. This App uses time-stamped photos, GPS locations, and facial recognition to check that the quarantines are staying safe at home.[14] The main advantage of this App is its ability to request the quarantines to take a selfie in their air-marked location on a routine basis. Non-receipt of such photos within the stipulated time is viewed seriously and stringent actions are initiated against such persons. This App also allows the users to send requests for essential commodities including foods by avoiding stepping out during the quarantine days. This model works well only for less populating countries with law-abiding citizens.

1.2.7 RUSSIA

The Russian Federation government developed a system to trace the contacts of a person with an international travel history and infected people by using mobile phone geolocation data. In this system, the users are allowed to send data whenever they come into contact with someone who is infected and processed information is forwarded to respective special regional headquarters.[15] The cellular operators are freely allowed to gather geolocation of the mobile phone app users. The technology-based on facial recognition is used to monitor the safe quarantine status of infected patients in Moscow city of Russia.

1.2.8 SINGAPORE

The Republic of Singapore Government released a mobile App called "Traced together" on March 21, 2020, to help in contact tracing.[16] This App is used to track the persons by using the proximity and the timestamps using Bluetooth technology. Whenever the need arises, this information can be used to identify the close contacts based on the proximity and duration of encounters between the infected and quarantined persons with the others. During the quarantine period, if a person is diagnosed with the COVID-19 virus, the government bodies access the data to identify their close contacts. The limitation of this Bluetooth-based App is that it only identifies the other nearby phones but not their locations.

1.2.9 SOUTH KOREA

The Center for Disease Control and Prevention of South Korea identified the travelers by using CCTV footage and tracked their location based on credit card records. After tracking, they were interviewed to diagnose the infection of COVID-19.[17] The government launched an App based on the mobile phone GPS data; such that it will send the message and trace the movements of the people who had a travel history from COVID-19 affected countries. Further, the App is also used to track the movement of quarantined people and also to collect data on the persons with the symptoms. The primary challenge of using this App is its working effectiveness relies upon GPS data. Either, due to unavoidable circumstances, or sometimes intentionally, the people being monitored may deviate from the quarantine without carrying their mobile phones and, in such cases, contact tracing becomes so difficult.

1.2.10 TAIWAN

The Taiwan integrated the National Health Insurance database for big-data analytics.[18] After identifying the cases with travel history from COVID-19 affected countries, real-time messages were generated based on clinical symptoms during diagnosis. On February 14, 2020, the entry quarantine system was launched and the travelers were made to furnish the health declaration form online by using a QR code. The persons with low risk were sent short messages on their phones for faster immigration clearance. The persons with high risk were quarantined and tracked through their mobile phones to ensure a safe stay during the stipulated period.

1.2.11 THAILAND

As a good initiative, Thailand's National Broadcasting and Telecommunication (NBTC) has launched a monitoring App accessed through the free sim cards provided to each and every international traveler on arrival.[19] Such people are advised to use the sim cards supplied and mobile App compulsorily during the quarantine period. The App is used to track the location of the users for 14 days in order to ensure their safe quarantine status. In case, if the people move from the quarantine location, the App will alert the authorities. However, the limitation of this App is its inability to trace the location of other people encountered with the quarantined person.

1.2.12 THE UNITED KINGDOM

The United Kingdom (UK) National Health Service (NHS) has developed a contact tracking App for monitoring travelers and infected patients.[20] The contact tracking App is based on the opt-in-basis and this is used to alert their neighbors about the new cases. This App allows the users to give inputs about their own symptoms that enable the confirmation of infection status through proper testing. British Telecom (BT), a UK-based mobile operator helps the government to monitor the quarantined persons by using the phone location and data usage. But the limitation is that anonymized data will be shared and updated information pertaining to the geographical movement of users will be delayed by 12–24 h.

1.2.13 THE UNITED STATES

The major tech firms such as Google and Facebook are in talks with the United States (US) Government to use the people's location data to combat the rapid spreading of COVID-19. These companies have proposed the use of AI and data analytics tools to estimate the chance of transmissions between the different regions.[21] The shared, aggregated and anonymized data are used to track the quarantines and to take the measures accordingly.

Mobile apps, Bluetooth technology, mobile data-based GPS service, facial recognition, timestamps, and wristbands are the various techniques used by the different countries for monitoring and controlling of COVID-19 spread. However, these technologies require active participation from the users. Furthermore, network capacity, bandwidth, device capacity, compatibility, battery life, lack of GPS accuracy in indoor environments, human interest, and commitments are the constraints for these technology-assisted monitoring.

1.3 DESIGN OF PROPOSED MONITORING SYSTEM

Based on the understanding of the requirements and analysis, the proposed system comprises two sections namely transmitter and receiver as illustrated in Figure 1.1.

The transmitter section contains radio frequency (RF) microchip (end-user equipment), GSM module, and satellite navigation system (GNSS) modules. These modules are used to transmit the exact geolocation information of the quarantined persons to the monitoring station. The miniaturized bio-microchip shall be implanted in the quarantined person's body at the time of entry. This RF microchip is a passive battery-less device that allows very short read within a defined communication range by using GSM/GNSS modules to ensure the safe status of quarantine in the specific location. In case, if a quarantined person moves from the earmarked location, RF microchip could not be read by the GSM/GNSS module, then the system will send alert messages to the authorities while alarming the quarantines. The location information will be sent to the monitoring station at the receiver side via GSM/GNSS module. This GPS and GNSS navigation module facilitates the storing and transmitting of the location information periodically for effective tracking. The monitoring-end terminal is used to capture the message pertaining to location information and display it on the electronic map for further processing.

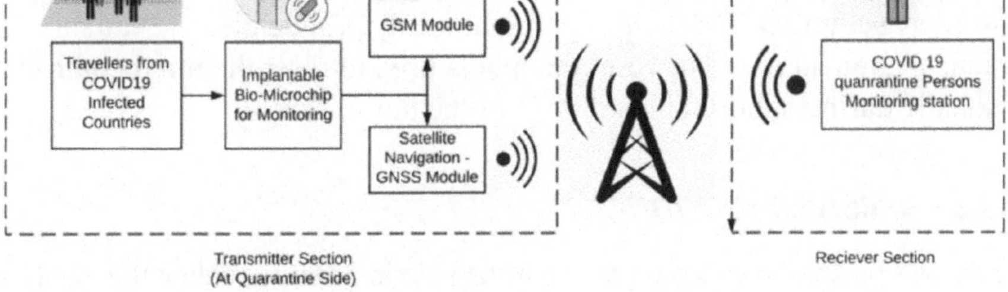

FIGURE 1.1 Proposed system design for monitoring quarantined person.

The overall system design are explained below.

1.3.1 TRANSMITTER SECTION

The transmitter side hardware design consists of bio-microchip, GSM module, and GNSS satellite communication modules.

The GNSS satellite navigation module is responsible for transmitting signals to the monitoring stations via satellites for tracking and navigation purposes in order to locate the quarantined person. The satellite navigation system module (UM220-III N) designed and developed by the GNSS SoC company consists of both GPS and navigation modules. The main advantages of this smaller size module are its low power consumption, high integration, and good compatibility.

The GSM module is responsible for sending the location messages through the SMC center to the monitoring station. The SIM900A GSM module uses surface mount technologies (SMT) packages and dual-frequency (GSM and GPRS). The GSM is connected via single-chip computer technology for realizing the GSM module controls. The serial-communication port is enabled for a further communication interface between the computer and navigational module. The microcontroller power module is used to connect the system via universal serial bus (USB) to DC 5.0 interfacing module.

The bio-microchip is a perfect MEMS-based computer technology that contains central processor functions, memory devices, input/output ports, interrupts, timers, and silicon pieces by using large-scale-integrated circuits. The microchip consists of a RF module to transmit and receive radio signals with GSM/GNSS navigation module in order to ensure the safe quarantine status of a person. On this basis, the geographic position coordinates will be processed and forwarded as a text message via GSM modules to the monitoring center by using serial communication modules. Due to the design impairments, unique manufacturing technology, and hardware design strategies, the single-chip yields ultra-running speed with comparatively low power consumption. This chip is portable, easily producible, and suitable for practical applications. The reset module is used to reset the microcontroller manually during abnormal conditions of the bio-microchip.

1.3.2 RECEIVER SECTION

Android App shall be used on the receiver side in order to realize the location details transmitted. The CRT monitor module is used to observe the current location of the quarantines and to keep tracking their movements based on the latitude and longitude positions. For simplicity and ease of understanding, such information is presented graphically by using the electronic map. Further, the voice playback module is used to broadcast the alarm to the local monitoring authority for immediate action.

1.3.3 SOFTWARE REALIZATION OF THE PROPOSED SYSTEM

The software of the transmitter side terminal equipment consists of the main program, data receiving and processing program, SMS sending program in the GSM module, and serial ports program. The proposed monitoring system program flow is illustrated in Figure 1.2.

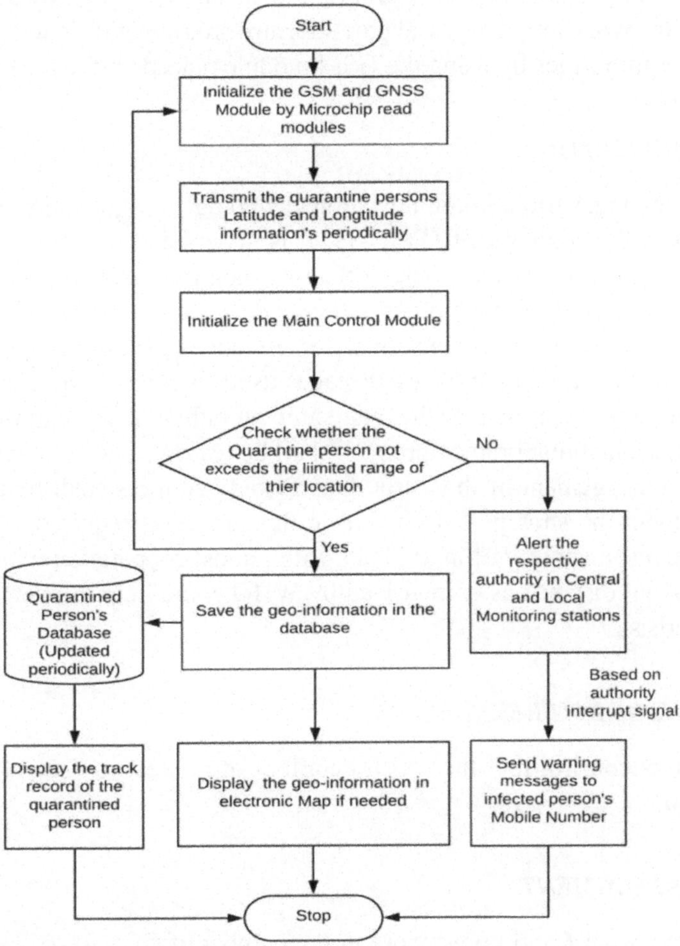

FIGURE 1.2 Program flow of the designed system.

The main program initializes all the modules based on the workflow. After initializing, the location information of the quarantined person is sent to the monitoring center via GNSS or GSM module. Such transmitted data consist of both the latitude and longitude information from time to time. Whenever the information exceeds a specific threshold limit (i.e., the undesirable movement of a quarantined person), the system will send alert messages to the local authorities and also send pop-up messages to the monitoring station. The distance (d) is calculated using the latitude and longitude information received from the GSM/GPSS modules. The program of the monitoring control unit (MCU) runs both the serial port subroutine, and the serial port interrupts the program to execute the stage-by-stage task based on the workflow. The serial port interrupt program will send data to the respective authorities by using the GSM module placed on the receiver side.

1.4 CONCLUSION

Upon reviewing, various communication technologies employed by different countries in the combat of COVID-19, it is understood that a combination of technological approaches helps effective monitoring of the spread. The implantation of bio-microchip and integrated use of GSM and GNSS navigation module shall be thought over for effective tracking of quarantines and their contact tracing in the early stages itself to prevent rapid spreading. By considering the impact of this pandemic on public health and the global economy, such an integrated approach shall be reviewed and considered for the early management of the crisis. Unaffected countries shall be prepared to implement the strategies discussed in this article as required. However, collective responsibility of individuals with utmost cooperation while maintaining social distance as emphasized by WHO could help to overcome the present crisis.

CONFLICT OF INTEREST

The authors confirm that there is no conflict of interest to declare for this publication.

ACKNOWLEDGMENT

We salute all the frontline warriors in the combat of COVID-19 especially the doctors, paramedical staff, health workers, and other frontline officials who are working relentlessly by risking their lives and WHO for their noble services.

KEYWORDS

- **COVID-19 outbreak**
- **quarantine**
- **tracking**
- **communication technologies**
- **bio-microchip**
- **GSM**
- **GNSS**
- **contact tracing**

REFERENCES

1. World Health Organization. WHO Director-General's statement on IHR Emergency Committee on Novel Coronavirus (2019-nCoV). Press Release [Online], 2020. https://www.who.int/dg/speeches/detail/who-director-general-s-statement-on-ihr-emergency-committee-on-novel-coronavirus-(2019-ncov).
2. World Health Organization. Updated WHO Recommendations for International Traffic in Relation to the COVID-19 Outbreak. COVID-19 Travel Advice [Online], 2020. https://www.who.int/news-room/articles-detail/updated-who-recommendations-for-international-traffic-in-relation-to-covid-19-outbreak
3. World Health Organization. Key Considerations for Repatriation and Quarantine of Travellers in Relation to the Outbreak of Novel Coronavirus 2019-nCoV. COVID-19 Travel Advice [Online], 2020. https://www.who.int/news-room/articles-detail/key-considerations-for-repatriation-and-quarantine-of-travellers-in-relation-to-the-outbreak-of-novel-coronavirus-2019-ncov
4. World Health Organization. Novel Coronavirus (2019-nCoV). SITUATION REPORT – 1 2020. https://www.who.int/docs/default-source/coronaviruse/situation-reports/20200121-sitrep-1-2019-ncov.pdf?sfvrsn=20a99c10_4
5. Xu, B.; et al. Epidemiological Data from the COVID-19 Outbreak, Real-Time Case Information. Sci. Data 2020, 7, 1.
6. World Health Organization. Considerations for Quarantine of Individuals in the Context of Containment for Coronavirus Disease (COVID-19). Interim Guidance [Online], 2020. https://www.who.int/publications-detail/considerations-for-quarantine-of-individuals-in-the-context-of-containment-for-coronavirus-disease-(covid-19)

7. Kamel Boulos, M. N.; Geraghty, E. M. Geographical Tracking and Mapping of Coronavirus Disease COVID-19/Severe Acute Respiratory Syndrome Coronavirus 2 (SARS-CoV-2) Epidemic and Associated Events Around the World: How 21st Century GIS Technologies are Supporting the Global Fight Against Outbreaks and Epidemics. Int. J. Health Geogr. 2020, 19, 1.
8. Federal Ministry of Transport and Digital Infrastructure. Measures Adopted by the Federal Ministry of Transport and Digital Infrastructure. Press Release [Online], 2020. https://www.bmvi.de/SharedDocs/EN/Articles/K/Corona/corona-measures.html
9. The Government of the Hong Kong Special Administrative Region. Inbound Travellers—Points to Note for Home Quarantine for Inbound Travellers. Press Release [Online], 2020. https://www.coronavirus.gov.hk/eng/inbound-travel.html
10. Government of India. Information about Arrogya Setu App and many more—COVID-19. Press Release [Online], 2020. https://mhrd.gov.in/sites/upload_files/mhrd/files/Arogya_2.pdf
11. Government of Tamil Nadu, Health and Family Welfare Department. COVID-19 Dashboard: Whom to Contact. Press Release [Online], 2020. https://stopcorona.tn.gov.in/about-coronavirus/
12. Tomar, A.; Gupta, N. Prediction for the Spread of COVID-19 in India and Effectiveness of Preventive Measures. Sci. Total Environ. 2020, 728, 138762.
13. Tidy, J. Coronavirus: Israel Enables Emergency Spy Powers. Article [Online], 2020. https://www.bbc.com/news/technology-51930681
14. Website of the Republic of Poland. Coronavirus: Information and Recommendations: Travel. Press Release [Online], 2020. https://www.gov.pl/web/coronavirus/travel
15. Government House, Moscow. Mikhail Mishustin Approves Measures to Prevent Spread of Novel Coronavirus in Russia. Press Release [Online], 2020. http://government.ru/en/news/39152/
16. A Singapore Government Agency Website. Help Speed up Contact Tracing with TraceTogether: A New App Uses a Community-Driven Approach to Identify Close Contacts of Users. Press Release [Online], 2020. https://www.gov.sg/article/help-speed-up-contact-tracing-with-tracetogether
17. Park, O.; et al. Contact Transmission of Covid-19 in South Korea: Novel Investigation Techniques for Tracing Contacts. Osong Public Health Res. Perspect. 2020, 1, 60–63.
18. Chen, F.; et al. Big Data Integration and Analytics to Prevent a Potential Hospital Outbreak of COVID-19 in Taiwan. J. Microbiol. Immunol. Infect. 2020, 4, 60–69.
19. DDC, Thailand. Measures to Control COVID-19. Press Release [Online], 2020. https://ddc.moph.go.th/viralpneumonia/eng/file/main/Measures_for_traverlers_from_aboard_v5.pdf
20. Mayor, S. Covid-19: Researchers Launch App to Track Spread of Symptoms in the UK. BMJ Clin. Res. Ed. 2020,; 368, m1263.
21. Vaishya, R.; Javaid, M.; Khan, I. H.; Haleem, A. Artificial Intelligence (AI) Applications for COVID-19 Pandemic. Diabetes Metab. Syndr. 2020, 14 (4), 337–339.

CHAPTER 2

Mobile Application Technology Innovation Design Learning Model Using System Management Learning Approach as a Covid-19 Solution at the National Development University "Veteran" Yogyakarta

PARYATI,[1] SALAHDDINE KRIT,[2] and AHMED A. ELNGAR[3]

[1]*ST., M.KOM (Doctoral Candidate in UITM Malaysia), Lecture University Pembangunan National "Veteran" Yogyakarta, Indonesia*

[2]*Faculty of Sciences, University Ibn Zohr, Agadir, Morocco*

[3]*Faculty of Computer & Artificial Intelligence, Beni-Suef University, Beni-Suef City, Egypt*

ABSTRACT

In the world of education, innovations in designing the learning models using mobile technology continue to grow rapidly. One of them is a learning management system, which is a learning model. In addition, we use a variety of strategies and patterns, as well as learning system thresholds, especially when dealing with the COVID-19 pandemic. The e-learning system is a digital media-based learning model. Mobile learning is a form of learning model that uses communication technology devices. This learning system was developed with the goal of creating mobile learning application software. The advantage of this learning application software is that it provides you with materials that you can access at any time. Material visualization is

Global Digital Transformation and the Covid-19 Pandemic.
Rajdeep Chowdhury, K. Martin Sagayam, Neyara Radwan, Ahmad A. Elngar (Eds.)

very interesting. Mobile learning is a learning model that uses wearable technology. With various advantages, mobile learning is an alternative learning model design that can increase the efficiency and effectiveness of the teaching and learning process. The result of making this application program is to produce a mobile learning model and a mobile learning application that applies the concept of learning anywhere and anytime and is user-friendly. The purpose of this learning model design is to develop a mobile learning-based learning design model using a learning management system develop a mobile service which is a web-based mobile service application. The stages of manufacturing are making mobile learning application software that contains learning materials and creating a mobile server model with a web-based mobile service application. The long-term goal is to produce mobile learning model designs and mobile learning applications that can be accessed anywhere and anytime and produce learning concepts that are fun, effective, and efficient in terms of time and energy as well as operational costs. The specific objectives are developing a learning design model, based on mobile learning using a learning management system, and developing a mobile service which is a web-based mobile service application. The application creation phase is to create an application program focused on developing mobile learning models. This involves creating a mobile learning application software that already contains some learning material content and then developing a design model for the learning application, developed into a mobile server with web-based applications for mobile services.

CHAPTER SUMMARY

Introduction of information and communication technology (ICT) in educational institutions is now essential as ICT use can be an indicator of an institution's success. A significant number of faculty are taking advantage of technological advances, using the Internet as online learning or what we usually hear about online learning. A new trend in today's e-learning world is known as mobile learning. Accessing online learning systems using portable media such as smartphones, iPhones, PCs, and tablets is being discussed and used in developed and developing countries such as the United States, especially in Indonesia. Using mobile learning to support the teaching and learning process can make teaching and learning activities more flexible. The use of mobile phone technology has traditionally been used not only as a means of communication and entertainment but also as a learning medium.

This is evident from several studies using mobile phone technology as a learning medium. The method used in this study was the R&D method, and the Dick and Carey model was chosen for the development design.[6]

Since the primary output of research is software, the media development process is equipped with special software development methodologies using the waterfall model software development life cycle methodology. The results of this research, once implemented, are expected to be commercialized in the form of mobile learning application software. Mobile learning application software offers the advantage of always-on-demand availability and visualization of interesting material. Mobile learning is a model of learning that takes place between locations and environments using technology that is easy to carry when the learner has a mobile phone. Due to its diverse possibilities and advantages, mobile learning has the potential to become an alternative learning resource that can increase the efficiency and effectiveness of teaching and learning processes.

2.1 INTRODUCTION

Learning is a need for self-generation of life that strives for itself, as humans are born with the urge to live their lives. Humans achieve independence and adapt to various environmental changes.[5] Various learning methods continue to be developed today to achieve an effective and efficient learning process that is applicable. The implementation of information and communication technology (ICT), in educational institutions is now a must, because the application of ICT can be an indicator of the success of an educational institution. Not a few lecturers take advantage of technological advances, by using the internet as online learning, or what we call online learning. Here are some new trends in the world of e-learning today. It is Indonesia that uses mobile learning and portable media such as smartphones, iPhones, PCs, and tablets to access online learning systems used in developed and other developing countries such as the USA. Supporting this teaching and learning process using mobile learning can increase the flexibility of teaching and learning activities. The use of mobile phone technology is not only focused as a means of communication and entertainment, but also as a learning medium. This is evident from several studies using mobile technology, the phone, as a learning medium[9] uses mobile phone technology as a learning medium for students at Novi His Sad University in Zrenjanin, Serbia. In addition, mobile phone technology is also used in the world of education. Department of Computer Science Education Undiksha[8] conducted a study to develop a mobile phone-based learning design on SQL teaching materials for advanced database courses. In this study, a mobile phone-based learning medium is created as a means of learning SQL material. The reaction of the students who were taught mobile phones was positive.

The use of ICT in education continues to evolve with different strategies and patterns, basically as forms of learning using electronic devices and digital media, e-learning systems, and special forms of mobile learning can be divided into learning using mobile communication devices and technology. The very high level of sophistication of mobile devices, their relative ease of use, and their increasingly affordable prices compared to personal computers are the driving factors that expand the use and application potential of mobile learning is. It is increasingly emerging as a new learning trend shaping learning paradigms that can be done anytime, anywhere. This case study was conducted at University Pembangunan National "Veteran" Yogyakarta, which has 5 faculties and 27 departments from undergraduate, master's, and doctoral levels. The majors at this university include undergraduate level information systems (S1), Department of Informatics Engineering undergraduate level (S1), economics major (S2), mining department (S3), and others.

This case study will be conducted in the Programming Algorithm course in the Department of Informatics Engineering. The reason for choosing the programming algorithm course is that it is a prerequisite course for new students majoring in Informatics Engineering, which is at University Pembangunan National "Veteran" Yogyakarta. Furthermore, the reason for choosing the informatics engineering major is that the ratio of the highest number of students from all majors at University Pembangunan National "Veteran" Yogyakarta is in the Department of Informatics Engineering. The use of mobile learning applications at University Pembangunan National "Veteran" Yogyakarta until the 2020/2021 academic year, no lecturers or teaching staff have used it. This has become a special motivation for me, as a teaching staff at University Pembangunan National "Veteran" Yogyakarta to make this learning model application.

2.2 LITERATURE REVIEW

TABLE 2.1 Literature Review.

No	Name	Title	Profit	Loss
1.	Zoran Vucetic et al.[26]	*Mobile School Service.*	Mobile phone technology is used as a learning medium for students at Novi Sad University in Zrenjanin, Serbia.	Students cannot be guided face-to-face with their lecturers if there are difficulties in the material being taught.
2.	Wirawan[25]	Development of Mobile Phone-Based Learning Design on SQL Materials in Informatics Engineering Undiksha.	Produce a mobile phone-based learning media as a means ofLearning for SQL material.	Students are easier to understand the material given offline because they can ask questions directly about the material presented.
3.	Paryati et al., 2021	Application of the Mobile Technology Learning Design Model Using Learning Management System on UPN Campus During COVID-19 Pandemic.	Produce a Mobile Learning model and Mobile Learning application that applies the concept of learning anywhere and anytime as well as the concept of fun learning.	Students are rarely able to ask for material, and direct guidance from the lecturer, if there are difficulties.
4.	Restu Lusiana, Tri Andar[18]	Guide Inquiry Learning Model in Structure Learning Algebra I.	The learning model that has been obtained is declared valid by the validators.	Students have difficulty understanding complex calculation material.
5.	MM Roro Forijati[19]	Research-based Learning Model in Microeconomics 3.	The learning design meets the criteria, namely effective student activities.	Students find it difficult to ask instructors with complex materials because they do not meet in person.
6.	N Setiawan[15]	Application of Creative–Critical Learning Model in Business Research Methodology Course in Accounting Study Program Students of Fise-Uny.	The ability of lecturers to manage the learning model is good.	Students have not been able to achieve the standard of completeness that is set such as the courses given by the lecturer.

TABLE 2.1 *(Continued)*

No	Name	Title	Profit	Loss
7.	M Afand[13]	Models and Learning Methods of Sultan Agung Islamic University.	The ability of students to understand the material is good.	Students have not been able to meet the target of understanding the courses given by the lecturer.
8.	B Lestari[5]	Quality Improvement Learning with Model Learning CooperativeLearning.	The ability to test student learning outcomes is valid.	Students have not been able to obtain complete material quality standards.
9.	Hasbullah and Maman Somantri[9]	Development of E-Learning Model to Improve the Quality of Students' Processes and Learning Outcomes in Energy and Conversion Courses.	The ability to understand the level of material reliably is good.	Students cannot meet the target of optimal understanding of the material.
10.	B Azmy[6]	Model Inquiry LSLC Student Response.	Student responses are happy and positive. Students are able to respond to the application of the inquiry learning model in learning courses.	Students have difficulty in understanding the explanation of complex calculation formulas in the courses given by the lecturer.
11.	H. Kurniawan[10]	Mobile Learning Media Using Android.	Very supportive of efficient learning, Android-based mobile learning methods are very effective in teaching and learning activities.	Not every student has an android phone.
12.	Rif'ati Dina Handayani[16]	Development of Mobile-based Electronic Teaching Materials Learning on Wave Lectures.	Mobile learning that has been developed has a score of 3.8 and is included in the valid criteria.	Semester 3 students still have difficulty understanding the wave material.
13.	Nur Hidayah Hanifah et al[14]	Mobile Learning in Mi/Sd Learning Strategy Courses: Development Research.	Useful for students to practice teaching and understanding course material on MI/SD learning strategies.	Students lack control and lack of concentration during lectures.

2.3 EQUALITY

In the making of this book chapter entitled Learning Model Design Innovation using Learning Management Systems with Mobile Technology at the University Pembangunan National "Veteran" Yogyakarta during the COVID-19 Pandemic, there are similarities between the discussion and the literature review mentioned above. The similarities in the discussion of the theme are about learning models using mobile technology and e-learning as well as mobile learning.

2.4 RESEARCH PURPOSES

The objectives of this study are: To develop a learning design model based on mobile learning using a learning management system. Developing a mobile service which is a web-based mobile service application.

2.5 RESEARCH URGENCY

This research was conducted to create products in the form of mobile learning application software. Mobile learning application software benefits from the availability of ready-to-call materials and the visualization of interesting materials. Mobile learning is a learning model that takes place between locations or environments using technology that is easy to carry when the learner is in the mobile and mobile phone state. Due to its diverse possibilities and benefits, mobile learning is becoming an alternative learning resource that can increase the efficiency and effectiveness of teaching and learning processes.

2.6 LEARNING MEDIA

The learning set by the lecturer includes teaching objectives, teaching materials, teaching methodology, and teaching assessment. Teaching materials are a set of scientific material consisting of facts, concepts, principles, and generalizations of science that comes from the curriculum and can support the achievement of teaching goals. The teaching methodology is the methods and techniques used by lecturers in interacting with students so that teaching materials reach students and students become master of the teaching objectives. In the methodology, there are two aspects that stand out the most, namely teaching methods and teaching media as teaching aids. Meanwhile, assessment is a tool to measure or determine the level of achievement of a teaching goal.[1] Learning patterns that utilize learning media as sources in addition to lecturers can be described as follows:

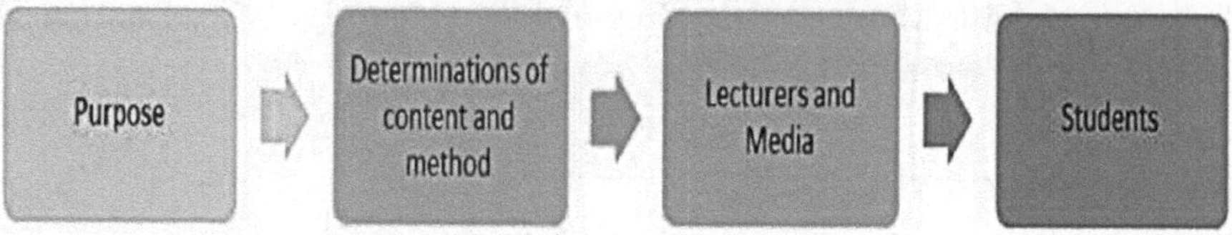

FIGURE 2.1 Media assisted learning pattern.

In learning practice, there is actually no rigid pattern between learning components. The complete combination pattern can be described as follows:

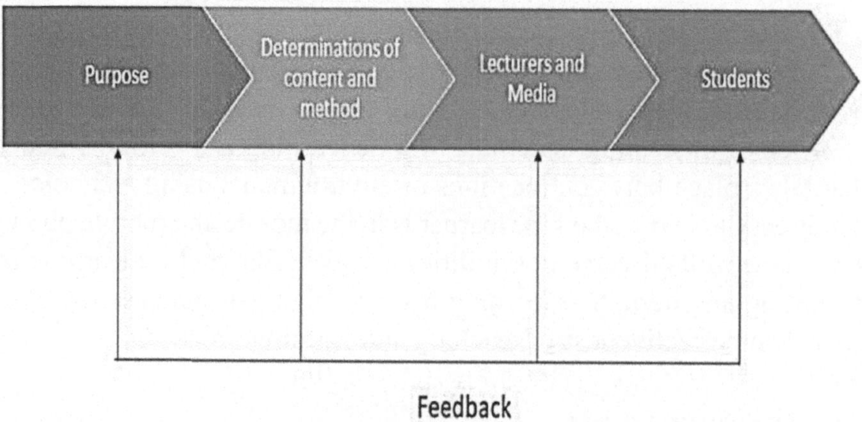

FIGURE 2.2 Combination patterns in learning.

Learning media are the means of communication in the teaching and learning process in the form of hardware and software that effectively and efficiently achieve the process and learning outcomes and facilitate the achievement of learning objectives. The purpose of using learning media is to enable students to create new things and use existing ones in other formats and variations that are useful in their lives. In this way, students can easily follow and comprehend the material that the instructor provides them.

2.7 MOBILE LEARNING

a. Basic Concepts

Mobile learning (m-learning) is learning using technology and mobile devices. In this case, these devices are PDAs, mobile phones, laptops, tablet PCs, and so on. Mobile learning allows users to access learning content anytime, anywhere, even when they aren't in a specific location at a specific time. This gives users access to educational content regardless of time and place. It implies that e-learning is a concept of distance learning using telecommunications and information technology.

b. Mobile Learning

Mobile learning is a learning model that takes place between locations or environments using technology that is easy for learners to carry around when using their mobile phones. Mobile learning, with its diverse possibilities and benefits, should become an alternative learning source in the future. This can improve the efficiency and effectiveness of the process and the learning outcomes of Indonesian students.[5]

c. Functions and Benefits of Mobile Learning

There are three functions of mobile learning in learning activities in the classroom (classroom instruction), namely as a supplement (optional), complementary (complement), or substitute (substitution). There are three alternative models of learning activities that students can choose, namely:

- completely face-to-face (conventional)
- partly face-to-face and partly through the internet
- completely via the internet.

d. Mobile Devices

A mobile device is a device that is easily carried everywhere, is light and can easily be carried in our hands. Mobile devices come in many varieties in terms of size, design, and layout, but they share very different characteristics from desktop systems.[4]

e. Mobile Device Application Platform

Some are familiar to us such as IOS, Android OS, BlackBerry OS, and so on.

FIGURE 2.3 Types of mobile.

f. M-Learning

The term mobile learning (m-learning) refers to the use of handheld and mobile information technology (IT) devices such as PDAs, mobile phones, laptops, and tablet PCs in education and learning.[2] Since M-learning is part of electronic learning (e-learning), it is also part of distance learning (d-learning).

FIGURE 2.4 Schematic of m-learning.

M-learning is unique because learners can access learning materials, guides, and learning applications anytime, anywhere. This can increase awareness of learning materials, make learning ubiquitous, and motivate learners for lifelong learning. In addition, m-learning offers opportunities for *ad-hoc* collaboration and informal interaction among learners compared to traditional learning. Some of the advantages of m-learning over other learning are:

1. Can be used anywhere at any time.
2. Most mobile devices are relatively cheaper than desktop PC.
3. The device size is smaller and lighter than a desktop PC.
4. Is estimated to be able to include more learners because m-learning utilizes technology commonly used in everyday life.

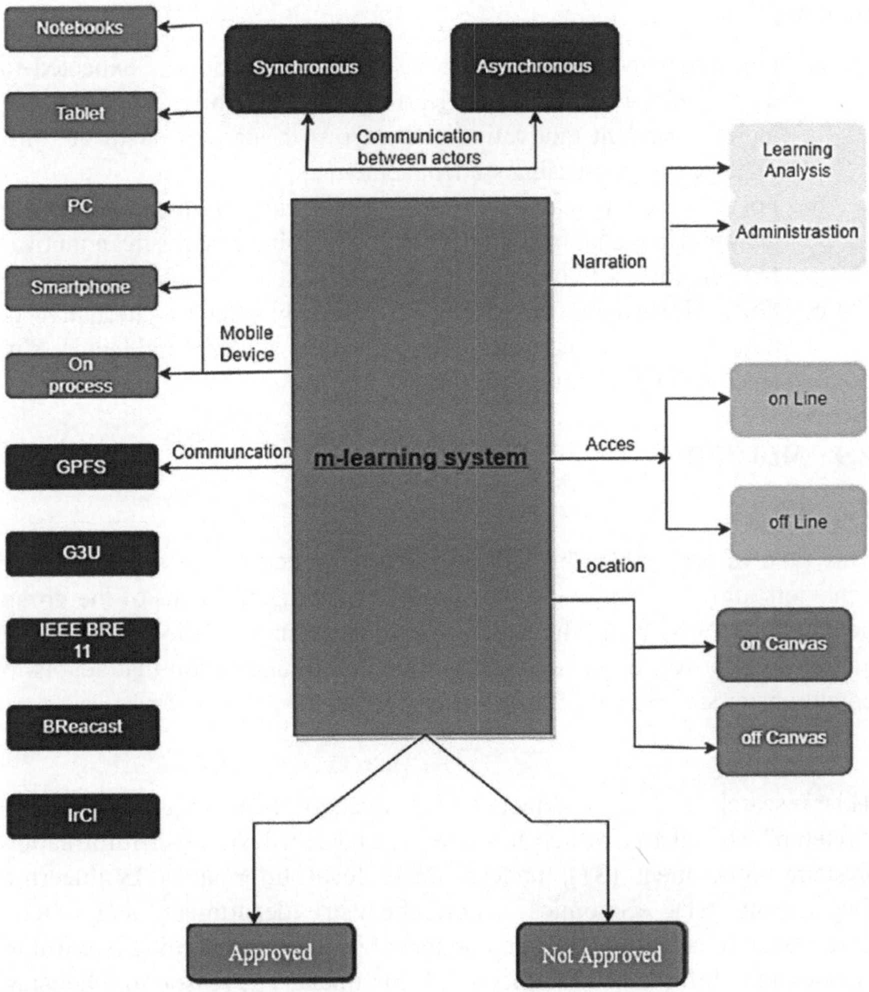

FIGURE 2.5 m-learning classification.

2.8 OBJECTIVES AND BENEFITS

Purpose

a To develop m-learning applications on mobile phones using Java technology.
b. To implement m-learning applications on cellphones so that they can run on different types and types of cellphones.
c. To develop an update system for web-based m-learning applications.

Benefits

a. For students, the m-learning-based learning model is expected to help students manage their learning independently. So that it can increase student motivation and interest in learning, with the ultimate goal of increasing student achievement.
b. For the teaching staff, the results of this study are expected to help lecturers in managing and developing learning designs in an innovative, creative, and interactive way.
c. For policymakers, the results of this study offer an alternative to new learning technologies, which can help students and lecturers in teaching and learning activities.

2.9 METHOD

Flow Chart

This type of research is behavioral research aimed at the application of behaviors aimed at improving the quality or solving problems of the group being studied, observing the success or consequences of the behavior and aiming to improve the behavior. It is inherited by the action that adapts to conditions or situation to achieve better results.

Location

This research will be conducted at University Pembangunan National "Veteran" Yogyakarta, which has one (1) undergraduate level Information Systems department (S1), undergraduate level Informatics Engineering Department (S1), economics major, chemistry department, and others. This research case study will be conducted in the Programming Algorithm course in the Informatics Engineering Department. The reason for choosing programming algorithm courses is that these courses are prerequisite courses for new students for all majors at University Pembangunan National

"Veteran" Yogyakarta. Furthermore, the reason for choosing the informatics engineering department was that the ratio of the number of students from all departments at University Pembangunan National "Veteran" Yogyakarta was high in the Department of Informatics.

Types of Methods

The research and development methods used in this chapter, where the development design chosen is to use the Dick and Carey Model quoted from.[7] Because the main output to be produced in research is in the form of software, the media development process will be equipped with a special software development method using the Software Development Life Cycle (SDLC) method with the Waterfall-based Model. Meanwhile, the specific targets of this study are (1) To develop a learning design model based on m-learning using a learning management system/LMS;[3] (2). Developing a mobile service which is a web-based mobile service application.

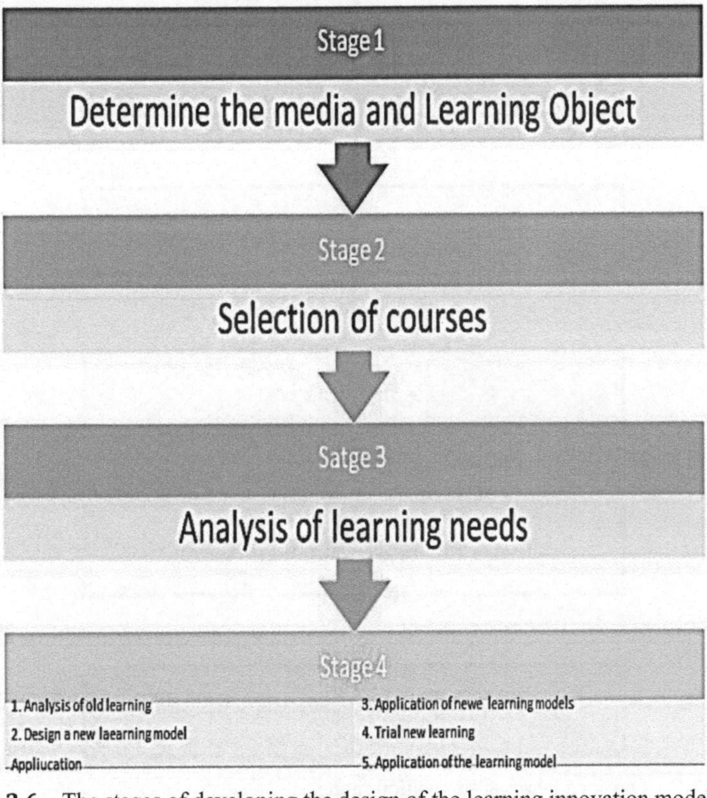

FIGURE 2.6 The stages of developing the design of the learning innovation model.

Since the media developed in this study produces the final product in the form of simulation program software, the third phase was followed by the fourth phase of development design that conforms to a software development methodology, the Software Development Life Cycle (SDLC) methodology. The stages follow waterfall model. The SDLC methodology is a methodology for analyzing and designing structured systems.

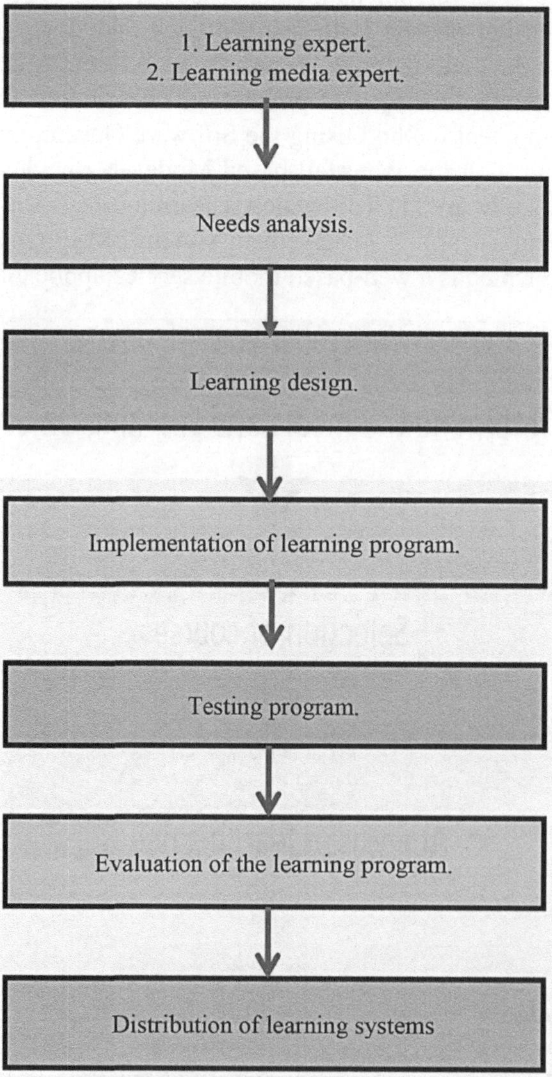

FIGURE 2.7 Design a learning innovation model.

The fifth phase is the final phase of the design of learning media development, in which software testing as learning media is performed. This includes media expert testing, content expert testing, small group testing, and field testing.

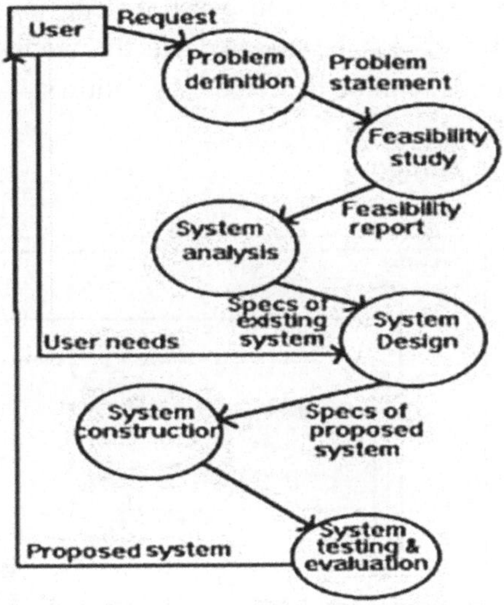

FIGURE 2.8 Mobile web application design model.

Variables

The independent variable in this study is the m-learning media developed in this study, while the dependent variable measured in the study is the student's response to the development and use of m-learning as a learning interaction aid.

Data Collection Techniques

The data collection of this research was carried out by observation, interview, and questionnaire documentation (questionnaire) to measure the likelihood of student learning outcomes.

Data Analysis and Interpretation

Quantitative data for measuring the success of learning media (mobile learning) using mobile phones and laptops were analyzed using descriptive statistical techniques. On the other hand, qualitative data from observations and interviews were analyzed using categories and coding techniques.

2.10 RESEARCH OUTCOMES

The output of the first phase of research is in the form of application tools/learning media application software on mobile phones which are then called m-learning applications, which can operate on various types of cellphones. Then the second stage output is a web-based m-learning application known as the mobile web which can be updated via the website. The m-learning application design model can be described[6] as follows:

a. First stage design model

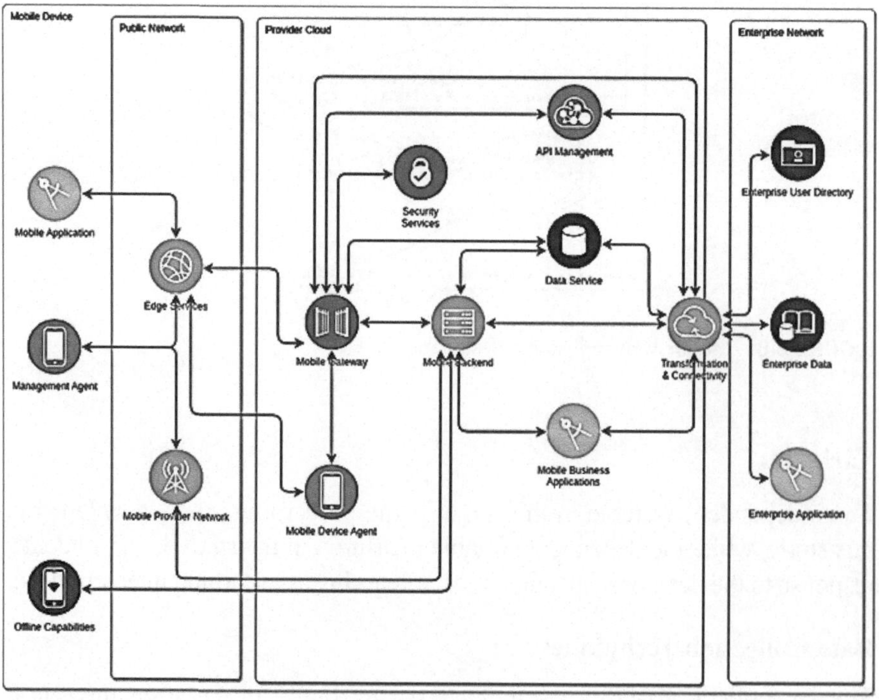

FIGURE 2.9 Design of the m-learning model.

In this first stage, it focuses on the use of mobile devices such as cellphones, smartphones, PDAs, and so on.

b. Second stage design model

The mobile web design model is described as follows:

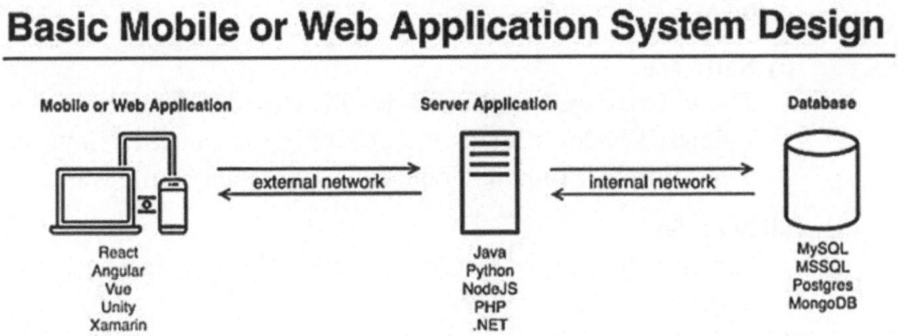

FIGURE 2.10 Mobile web application design model.

c. Development of a m-learning system

The development of the mobile system is directed at two sides, namely as follows:

- Developing a mobile service which is a web-based mobile service application on the server-side.
- Developing a mobile client, which is a mobile application on the client-side that can access mobile server services.

d. Outcome indicators

The indicators in developing this application are the successful development of a mobile learning application on a cellphone that can be operated on various types of cellphones and the building of a web server information system site for updating them-learning application.

e. Research facilities

The research facilities required are as follows:

(i) Hardware

1. Laptop specifications for editing purposes, with detailed specifications:

   Intel centrino Core 2 Duo processor (2.1 Ghz, 800 Mhz FSB),

   - 4 DDR memory.
   - 320 GB hard drive.
   - 512 MB video graphics

2. Canon MP 258 Printer

(ii) Software

The software specifications that will be used in this study are as follows: Windows XP as the operating system, Java Environment, Wireless Toolkit, Graphics Programming with Java.[9]

(iii) Services

Services are services that are used are 1 Gbps internet services.

2.11 RESULTS ACHIEVED

Overview of Learning Models at University Pembangunan National "Veteran" Yogyakarta.

Based on the results of observations, the learning activities used at University Pembangunan National "Veteran" Yogyakarta for all majors and levels are generally still lecture or conventional, especially, for some general basic courses. The vision and mission of the informatics engineering department are to produce computer graduates who master software engineering, computer network, and system design, computer graphics applications, web design, computer-aided design, information systems, database systems, multimedia, and programming language skills.

Learning Technology-Based on Mobile Learning

The method used in this research is a research and development method, with a Dick and Carey model development design, with the main output in the form of software. So that in the development process using m-learning-based learning media, the method used is equipped with a special software development method, namely the software development life cycle (SDLC) method with the waterfall-based model. The design of the research method development model is described as follows:

1. Determination of courses
2. Needs analysis is functional needs and non-functional needs :
   i. Operational can be accessed by clients from various browsers that support java script. Can be accessed on android & blackberry smartphones.
   ii. Security. There is the use of a password in the login form to distinguish the type of user, including their respective access rights.
   iii. The draft development process follows:

(a) Analysis of learning conditions

Based on the results of observations of the algorithm & programming course, the data obtained were analyzed using a qualitative description. The results of the descriptive qualitative analysis show the characteristics of the department informatics engineering learning model.

(b) Development steps:

1. The results of descriptive qualitative analysis for development prospects and opportunities.
2. The results of descriptive qualitative analysis for the development of learning technology.

(c) Measuring steps of learning outcomes

The assessment indicators that apply are in accordance with the applicable academic rules at University Pembangunan National "Veteran" Yogyakarta, the range of values applies.

(d) Mobile learning development design

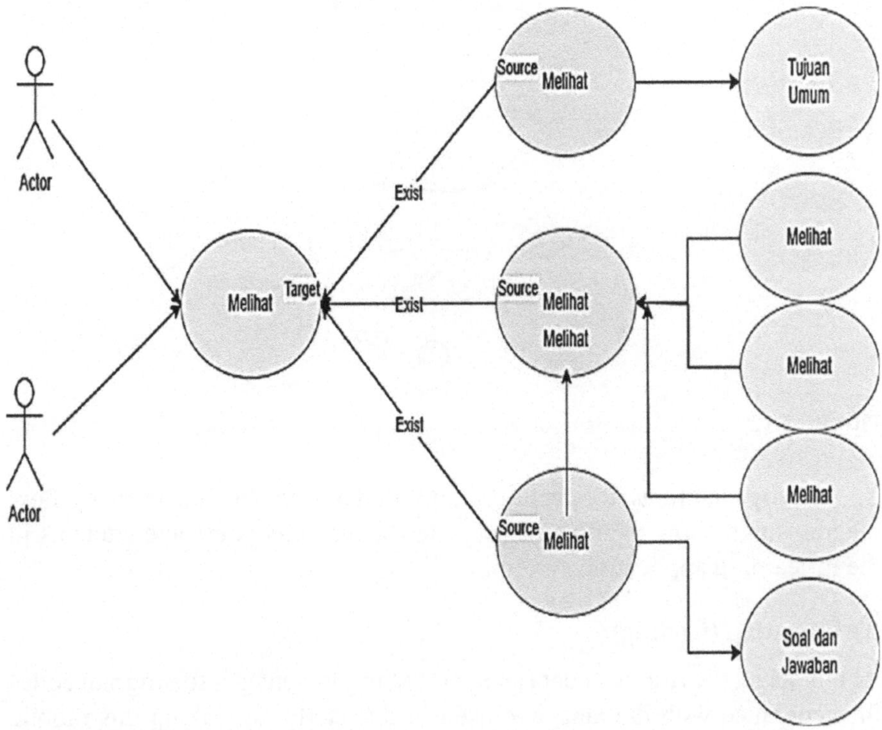

FIGURE 2.11 Use case diagram.

(e) Activity diagrams for lecturers and students

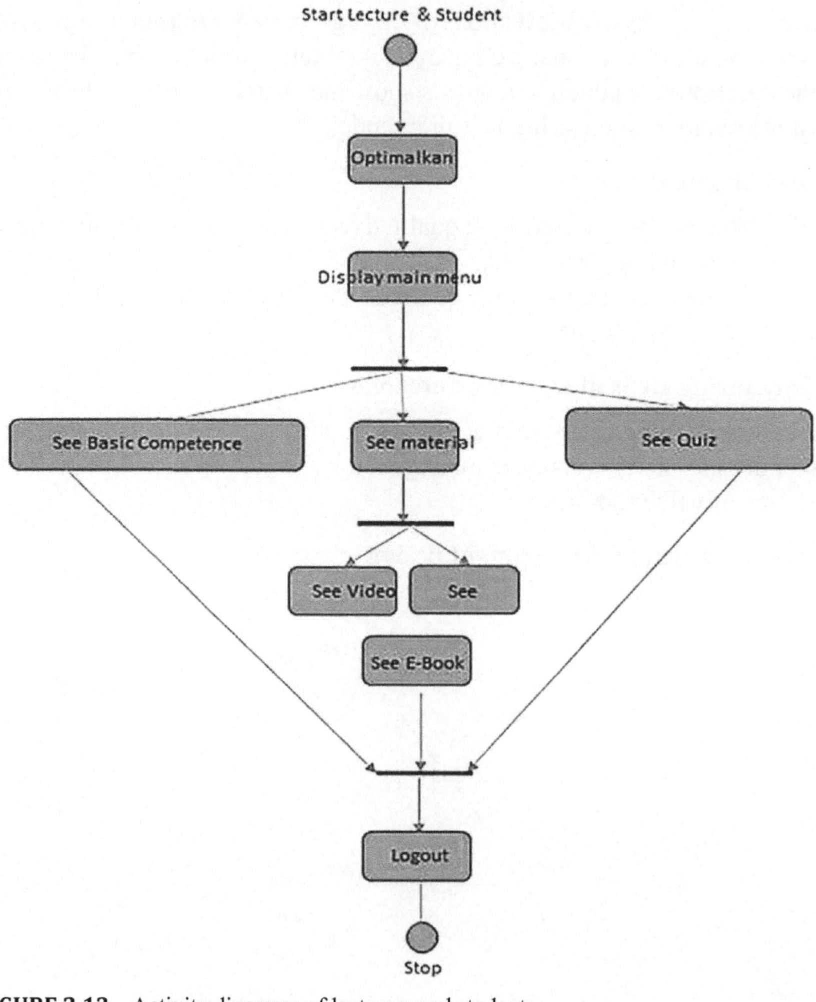

FIGURE 2.12 Activity diagrams of lecturers and students.

Activity diagrams for lecturers and students are in the form of flow diagrams that show the flow of activities between lecturers and students in the m-learning application system.

(f) Encoding (Coding)

At this stage, the results of data analysis are implemented into program codes in accordance with the language used in this study. In making this mobile learning program, we use the Java NETBEANS 7.2 language.

2.12 IMPLEMENTATION

After arriving at the stage of making m-learning software as described in the previous chapter, the target according to the next research methodology will focus on the following activities:

(a) Testing and implementation of mobile learning applications
(b) Mobile learning application testing
(c) Analysis of learning technology results based on mobile learning.

2.13 CONCLUSIONS

At the University of National Development "Veteran" Yogyakarta, they have not used mobile applications in the teaching and learning process. Most of the academic community, in the teaching and learning process still use mobile phones as a means of communication. Then a learning design model mobile application was made, using Net Beans software with J2ME. This learning design model mobile application that has been created can be run on several different operating systems: on m-learning devices such as Symbian operating system, Windows mobile operating system, Blackberry, and Android operating system. The innovation of mobile application development for learning design models on mobile phones uses Java technology. To implement a learning design model mobile application, mobile phones it can run on various types of mobile phones. The development of the learning design model system for this m-learning application is web-based. Making this application produces a mobile learning design model, and a mobile learning application that applies concepts learned anywhere and anytime as well as learning concepts that are fun and user friendly. The next development is a m-learning-based learning design model, using a learning management system, at the National Development University "Veteran" Yogyakarta. They succeeded in developing a mobile server, which is a web-based mobile service application. As well as successfully developing a mobile learning design model, namely by creating a mobile learning application software, which contains several learning material content. In addition to this, they also succeeded in developing a learning design model application, which was made in the first stage into a mobile server with a web-based mobile service application.

KEYWORDS

- **learning management system**
- **mobile technology**
- **learning model design**
- **e-learning**
- **mobile learning**

REFERENCES

1. Arsyad, A. *Technology Learning Media*; Raja Grafindo Persada: Jakarta, 2017.
2. Kadir, A. *Learning Management System*; Graha Media Publishing Agency: Yogyakarta, 2016.
3. Purnamasari, A. W. et al. *M-Learning Application on Android Platform*[Online], August 2014. Merpati Vol. 2, No. 2. https://ojs.unud.ac.id/index.php/merpati/article/download/17883/11614
4. Asmi, A. R. Development of Android Application-Based Mobile Learning Model on ASEAN History Course. *Candrasangkala: J. Educ. Hist.* [Online]2019, *5* (1), 30–40. https://jurnal.untirta.ac.id/index.php/Candrasangkala/article/download/4370/4284
5. Lestari, B. Quality Improvement Learning with Model Learning Cooperative Learning. *J. Econ. Educ.* [Online] 2008, *5* (2). https://media.neliti.com/media/publications/17221.
6. Azmy, B. *Model-Inquiry-LSLC-Student-Response* [Online]. https://ojs.unpkediri.ac.id/index.php/pinus/article/view/13437
7. Adinugraha, F. Project-Based Learning Model on the Eyes Lecture of Learning Media. *J. SAP* 2018, *3* (1). 9.
8. Herlinah, H. *Designing Mobile Learning as a Teaching Aid*; UNM Publishing Agency: Makasar, 2012.
9. Hasbullah & Maman Somantri, Development of E-Learning Learning Model to Improve the Quality of Students. *Processes and Learning Outcomes in Energy and Conversion Courses J. Res.* [Online] 2009, *10* (October 2).http://jurnal.upi.edu/penelitian-Pendidikan/view/556
10. Kurniawan, H. *Mobile Learning Media Using Android* [Online], 2017. http://jurnal.ubl.ac.id/index.php/explore/article/view/866.
11. Matlubah, H.; et al. Smart Phone-Based Mobile Learning Application Android as a Program Student Learning Source Study of Science Education, Sumenep Wiraraja University *J. Sci. Lanterns (Lens)* [Online]. 2016, *6* and II85. https://www.ejournalwiraraja.com/index.php/FKIP/article/view/290/248.

12. Hernawan, H. Use of HTML 5-Based Mobile Learning Applications to Improve Student Understanding in Courses of Microbiology. *PEDAGOGIA: J. Educ. Sci.* [Online] 2017, *124*. https://ejournal.upi.edu/index.php/pedagogia/article/download/8090/5114
13. Afandi, M. *Models and Learning Methods of Sultan Agung Islamic University* [Online], 2013. http://research.unissula.ac.id/file/publikasi/211313015/9230.
14. Hanifah, N. H. et al. Mobile Learning in Mi/Sd Learning Strategy Courses: Development Research. *Ta Dib* [Online] 2020, *23* (1), 119. https://www.researchgate.net/publication/342628228.
15. Setiawan, N. Application of Creative-Critical Learning Model in Business Research Methodology Course in Accounting Study Program Students of Fise-Uny. [Online] 2009. https://journal.uny.ac.id/index.php/jpakun/article/view/946.
16. Handayani, R. D. Development of Mobile-Based Electronic Teaching Materials Learning On Wave [Online]. 2015. https://repository.unej.ac.id/bitstream/handle/123456789/73664.
17. Rosa, A. S.; Shalahuddin, M. *J2ME Programming Quick Learning to Program Mobile Telecommunication Devices*; Modula Publisher: Bandung, 2018.
18. Lusiana, R.; Andar, T. Guide Inquiry Learning Model in Structure Learning Algebra I. *J. Math. Sci. Educ* [Online]. 2015. http://ejournal.unipma.ac.id/index.php/JEMS/article/view/245
19. Roro Forijati, M. M. Research Based Learning Model in Microeconomics 3 [Online]. 2019. https://jurnal.stkippgritulungagung.ac.id/index.php/jupeko/article/download/1035/537.
20. Semiawan, C. R. *Learning Foundation in Human Development, Center of Human Competency Development*; Elex Media Komputindo: Jakarta, 2017.
21. Santyasa. Research Methods and Module Development Theory. In *Paper Presented in a Training for High School, Vocational School Teachers*, January 14–16, in Klungkung district, Nusa Penida district, 2009.
22. Maudiarti, S. *The Application of e-Learning in Universities* [Online], 2020. https://media.neliti.com/media/publications/259453
23. Listyorini, T. Designing of Mobile Learning Courses of Android-Based Operating Systems. 2013. DOI: 10.24176/Simet.V3i1.8
24. Taufiq, A. *Graphics Programming with Java*; Andi: Yogyakarta, 2017.
25. Wirawan, I. M. A. In *Development Learning Media for Mobile Phone Based Materials Basic SQL Syntax in Subjects Advanced Database Case Study on IT Educational Programs Semester III*, Proceedings of the International Seminar on Information Technology and Education, Bridging ICT and Education, DIN EN ISO 9001: 2008, Cert. No. 01 100 086042, Faculty of Engineering, Padang State University, 2019.
26. Vucetic, Z.; et al. Mobile School Service (MSS). *Int. J. Interact. Mob. Technol.* 2018, *4*.
27. El-Sofany, H. F., & El-Haggar, N. (2020). The Effectiveness of Using Mobile Learning Techniques to Improve Learning Outcomes in Higher Education. *International Journal of Interactive Mobile Technologies* (iJIM), *14*(08), pp. 4–18. https://doi.org/10.3991/ijim.v14i08.13125

CHAPTER 3

A Detailed Study on the COVID-19 Outbreak in India and a Deep Learning-based Approach to Detect Lung Infection Using CT Images

J. ANDREW ONESIMU[1], R. JENNIFER EUNICE[2], LENNY CHRISTOPHER[3], and AHMED A. ELNGAR[4]

[1]*Department of Computer Science and Engineering, Manipal Institute of Technology, Manipal Academy of Higher Education, Manipal, Karnataka, India*

[2]*Department of Electronics and Communication Engineering, Karunya Institute of Technology and Sciences, Coimbatore, Tamil Nadu, India*

[3]*Department of Bioinformatics, Karunya Institute of Technology and Sciences, Coimbatore, Tamil Nadu, India*

[4]*Faculty of computers and Artificial Intelligence, Beni-Suef University, Beni-Suef City, Egypt*

ABSTRACT

On December 31, 2019, World Health Organization (WHO), China country office, was informed about new cases of pneumonia with unknown factors detected in Wuhan City hospital, Hubei province of China. Resources say that the outbreak of this epidemic novel coronavirus (nCoV) was said to have spread from an animal source to humans from Wuhan's Huanan seafood market, China. COVID-19 detected in China is closely related to the severe acute respiratory syndrome (SARS-CoV-1) virus and it is also named

Global Digital Transformation and the Covid-19 Pandemic.
Rajdeep Chowdhury, K. Martin Sagayam, Neyara Radwan, Ahmad A. Elngar (Eds.)

SARS-CoV-2 virus. While animals are believed to be the source of spread, the virus is now spread commonly from person to person and claimed thousands of lives across the globe. The onset of this symptom raged between December 8, 2019, to January 12, 2020, and within this period, a total of 198 cases were identified with nCoV in China. Out of these 198 cases, 136 cases were identified in Wuhan, 2 in Beijing, 1 in Guangdong, and 2 in Shanghai. During this period, nine exported cases were reported from different countries, such as one in Japan, two in Thailand, and one in South Korea. Of these reported cases, 43 were severe illnesses, 9 were said to be critical, and the remaining were stable under medical observation. On January 30, 2020, the first confirmed case of COVID-19 was diagnosed in a laboratory in Kerala, India, from an infected person's sample. This patient was a student returning from Wuhan. Day by day, the number of active cases and death rates was mounting tremendously creating a pandemic outbreak of the SARS-COVID-2 virus in most countries. This sudden outbreak of COVID-19 has caught many countries unprepared for treating the illness and they are struggling to find an antidote for the virulent novel coronavirus. Since COVID-19 is a disease with no immediate cure the first resort was early detection of the virus and isolation of the patients. RT-PCR kits were used as the primary method of diagnosing COVID-19. It has been observed that these tests produced results indicating the patient doesn't have a virus when they do. The rates of false negatives have ranged from 2% to 37% for RT-PCR kits whereas it is around 50% for the antigen tests. CT scans were used for a more accurate result reducing the inefficiency of the initial tests. The prices and need for chest X-rays and HRCTs have exponentially shot up due to the high demand for these tests. A CT or computerized tomography scan makes use of computers and rotational X-ray machines for creating cross-sectional images of the human body. The application of deep learning in the field of COVID-19 radiologic image processing reduces false-positive and negative errors in the detection and diagnosis of this disease and offers a unique opportunity to provide fast, cheap, and safe diagnostic services to patients. In our chapter, we aim to build a convolutional neural network to detect COVID in a CT scan. Extensive training of the network using our models has achieved a training accuracy of 96.04% and a testing accuracy of 95.17%.

3.1 INTRODUCTION

The word "Corona" was derived from observing the virus with a crown-like morphology under an electron microscope in 1968. This virus can infect many animal species like bat, rat, cat, pig, cow, chicken, turkey, and humans.[34] According to a study, the human coronaviruses in their earlier stages induced only a common cold among humans. In 2003, an outbreak of severe acute respiratory syndrome (SARS)[19] was reported in China, where the cause of this SARS was due to a new variety of human coronaviruses. This epidemic outbreak in China enlightened scientists regarding the evolution of the coronavirus and many scientific studies have been carried on since the outbreak.

The studies show that the coronaviruses are categorized into three different groups based on their genome sequence analysis. Group I include Human Corona Virus 229 (HCoV-229)[15] that causes respiratory infection. Other coronaviruses that falls under this category are porcine epidemic diarrhoea virus (PEDV), common among pigs resulting in respiratory and enteric infection; and feline infectious peritonitis virus (FIPV) among cats causing respiratory, enteric, and neurologic infection along with hepatitis. Human Corona Virus NL 63 (HCoV-NL63) causes respiratory infection and croup among humans. Group II category comprises SARS-CoV-1, CoV-HKU1, and HCoV-OC43. Human coronaviruses result in respiratory and enteric infections among humans. According to Woo et al.[36] Coronavirus HKU1 (CoV-HKU1) was identified from a 71-year-old patient with pneumonia. Group III category includes avian coronaviruses—turkey coronaviruses are most prevalent among chicken and turkey, causing respiratory infection, hepatitis, and enteric infection.

Coronavirus disease 2019, commonly known as COVID-19, is an infectious disease that has infected more than 103 million individuals all over the world and caused more than 2.8 million deaths. The virus causes mild-to-severe respiratory illness in people and can even result in death under extreme circumstances, especially if the patient has underlying medical complications.[12] COVID-19 has spread worldwide. The best preventive measures include wearing a mask, staying six feet apart, washing hands often, etc. The most common symptoms include fever, fatigue, dry cough, loss of appetite, body aches, and mucus. A huge obstacle that had been observed while trying to control the spread of this disease is the inefficiency as well as a shortage in several tests available to the general public. The tests that are being done now are majorly based on reverse transcription-polymerase chain reaction (RT-PCR), which is a laboratory test that can identify and recognize the virus which is responsible for causing COVID-19 in the patient. This RT-PCR test[7] is intended for detecting SARS-CoV-2 and takes several hours to obtain results, which is a long time compared to the rapid rate at which the COVID-19 disease is spreading in society. There is an urgent necessity to recognize the illness in the initial stage itself and ensure that the infected patient is immediately under quarantine because no specific drugs are known and available for COVID-19. A potential carrier who is awaiting results might interact with people around him thereby spreading germs and the virus unknowingly. Significant damage would already be done by the time the results are available and the carrier is put under quarantine.

In current times, there has been a worrisome shortage of RTPCR test kits all over the world, added to the extreme load that is being placed on laboratories and medical facilities. It has been observed that RT-PCR tests show low sensitivity. This is not satisfactory as the world is in a state of pandemics right now with increasing deaths and worsening conditions but no solution is being found. It has also been observed in a few COVID-19 cases that the infected persons are not being identified on time and are, therefore, not receiving the suitable treatment. There is also a risk of the infected person being assigned COVID-19 negative because of false-negative results which are more than plausible. To mitigate the inefficiency and shortage of existing number tests for COVID-19 detection, many efforts have been devoted to searching for alternative testing methods. Several studies have shown that computed tomography (CT) scans manifest clear radiological findings of COVID-19 patients and are promising in serving as a more efficient and accessible testing manner as CT devices are available and accessible across large areas and are capable of generating results at a fast and much more accurate pace.

ACT or computerized tomography scan makes use of computers and rotational X-ray machines for creating cross-sectional images of the human body.[3,6,11,20;,23,24] These images that have been generated provide more important detailed information than normal X-ray images. They go on to show soft tissues, blood vessels, and even bones in various parts of the body. Further, to alleviate the burden of medical professionals in reading CT scans, several works have developed deep learning methods that can automatically interpret CT images[23] and predict whether the CTs are positive for COVID-19. These CT images enable physicians and medical personnel to identify the internal structures in the human body and also see their shape,[30] density, size, as well as texture. Being significantly distinct from the X-Rays taken in the usual way, computerized tomography scans generate a set of slices from a particular part of the human body without overlaying the various body structures. Hence, we can say that these scans give a better and more detailed picture of the infected patient's condition than the conventional X-Ray images. This detailed information can be used to understand if there is a medical problem or not as well as comprehending the exact location where the problem had originated. For the reasons mentioned above, an increasing number of deep learning-based methodologies are being proposed and developed for COVID-19 screening in CT scans.

3.1.1 COVID-19

Novel coronavirus is a disease that causes severe common cold to severe respiratory infections among humans. This virus has its evolution from SARS and Middle East respiratory syndrome (MERS). The virus which has been spread among 210 countries as of this writing is declared a pandemic by WHO. The mean incubation period observed in travelers from Wuhan is estimated to be 6.4 days[8] and the incubation period ranges from 2.1 to 11.1 days. The studies provide valid evidence of symptoms among a family who was exposed to this virus by a traveler from Wuhan. The family developed symptoms of COVID-19 within 3–6 days of exposure. WHO has predicted an incubation period of 0–14 days. On January 22, 2020, the local government from Hubei Province, China, have reported that the observed incubation period among the patients is 27 days.[14] However, on February 10, 2020, WHO in a press meet declared that a very long incubation period may reflect double exposure.

3.1.2 TRANSMISSION MODES

The symptoms of COVID-19 in its prodromal phase has a fever, dry cough, dyspnoea, diarrhea, and sore throat.[8,28] This virus spreads from human to human via small respiratory droplets while coughing and sneezing, when in close contact with the infected person within 6 feet distance.[35] The risk also prevails when conjunctivas and mucosae are directly exposed to respiratory droplets of the infected person. Transmission of COVID-19 can also occur by touching objects or contaminated surfaces in the immediate environments of the infected persons. This transmission occurs not only by direct contact with the infected persons but also when the virus is transmitted through indirect contact of the infected persons and some recent studies say that airborne transmission of COVID-19 is also possible under certain circumstances during medical support.

3.1.2.1 SYMPTOMATIC TRANSMISSION

A symptomatic transmission is referred to transmission from an infected person who is experiencing virus symptoms.[17] Initial studies and reports stated that the primary transmission of the disease is possible through symptomatic persons to others when they come in direct contact with each other, through their respiratory droplets, or by touching the contaminated objects and surfaces. According to Liu et al.,[16] the virus is virulent during the early onset of symptoms and the people are more contagious during this period rather than the later stages of the disease. CT scan results of all symptomatic patients had multifocal ground-glass opacity, reduces lymphocyte counts, and increased C-Level protein.

3.1.2.2 PRE-SYMPTOMATIC TRANSMISSION

The incubation period of nCoV-19 is on average between 5 and 6 days. For some, the incubation period or the pre-symptomatic period may extend up to 14 days. During this incubation period, some persons before the onset of the symptoms become contagious to other persons and are prone to virus transmission. Recognizing the pre-symptomatic transmission is still a challenge as there is no such proper evidence for the reason for virus transmission.

3.1.2.3 ASYMPTOMATIC TRANSMISSION

Studies reveal that asymptotic carriers are latent sources of COVID-19. This refers to the transmission of the virus from a person who does not develop any symptoms of COVID-19. The following table explains the spread of COVID-19 through asymptotic carriers.

3.1.2.4 CASE STUDY

The above cases in Table 3.1 brief us on the details of delivery of infection from asymptotical carriers of COVID-19 in a familial cluster. Case 1 from Luzhou—China is said to have attended a family reunion dinner with her family members; case 2–5 between January 23–25, and together they all celebrated spring festival on January 30. Previously there were no cases of COVID-19 in Luzhou and case 1 had no contact with anyone returning from Wuhan except case 3/4/5. In this familial cluster of 5, cases 2 through 5 showed no symptoms of COVID-19 before the outbreak of case 1 which confirms that there must be some asymptotic carriers among cases 2 through 5.

Case 6 is a businessman from Germany who attended a meeting in Munich with a Chinese business partner between January 20 and 21. This Chinese business partner was a Shanghai resident who visited Germany between January 19 and 21 who had no symptoms of COVID-19. Case 6 developed symptoms of sore throats and chills on January 24, 2020, and recovers by January 27, 2020, and gets backs to his work routine, and meets his colleagues. Meanwhile, the Shanghai business partner developed symptoms of COVID-19 and informs regarding her illness to the office. Four other persons from this office were tested positive after January 27. These four patients were the ones who had a personal meeting with patient 1. In this case, the indexed patient from Shanghai is presumed to be an asymptotical carrier since she developed no symptoms during the incubation period.

Case 7 who lives in Wuhan is presumed asymptotic carrier of COVID-19 travels to Anyang and meets five of her relatives on January 10, 2020. These relatives were diagnosed positive with COVID-19 by January 13 where the indexed patient developed no symptoms until January 28. Case 8 attended a religious meeting in Delhi, India, and is a presumed asymptotic carrier of COVID-19 developed no symptoms but infected his wife aged 60 in Karnataka, India.[13]

TABLE 3.1 Asymptotic Transmission of COVID-19.

Case	Patients (male/female)	Age	Symptoms	Onset of symptom	Place/city	Reference
1	Woman	50	Fever, dizziness, cough, and shortness of breath	February 4, 2020	Luzhou, Sichuan province, China	Van Erum et al.[32]
2	Man	28	Fever	February 5, 2020		Van Erum et al.[32]
3	Woman	51	Pharyngalgia, diarrhea	February 5, 2020		
4	Man	50	Fever, cough	February 5, 2020		
5	Man	23	Fever, cough	February 5, 2020		
6	Man	33	Sore throats, chills, myalgias	January 24, 2020	Germany	Rothe et al.[28]
7	Woman	*20*	No symptoms	January 13, 2020	Anyang	Li et al.[14]
8	Man	> 60	No symptoms	April 6, 2020	Karnataka/ India	Ahmed et al.[2]

3.2 THE PANDEMIC COVID-19 IN INDIA

On January 30, 2020, the first case with COVID-19 symptoms emerged in India. Later, the travel history of the concerned person was tracked and found that the person was an Indian student returning from Wuhan, China. The laboratory reports tested positive and the patient was isolated in a special isolation ward in Kerala, India.[17] Between January 30 to February 3, 2020, three more cases tested positive for COVID-19 conducted in the laboratory of Kerala, India.[18,25,31] The Ministry of Health, Family, and Welfare (MoHFM) and cabinet secretary was assigned to closely monitor the novel coronavirus situation in India on daily basis.

By March 14, 2020, the active cases in India rose to 84 with 67 Indian nationals and 17 foreign nationals.[1] These cases were reported from different states such as Andhra Pradesh (1 case), Delhi (7 cases), Haryana (14 cases), Jammu & Kashmir (2 cases), Karnataka (6 cases), Kerala (19 cases), Ladakh (3 cases), Maharashtra (14 cases), Punjab (1 case), Rajasthan (3 cases), Telangana (1 case), Tamil Nadu (1 case), and Uttar Pradesh (12 cases). For the first time, the death of two patients due to COVID-19, a 76-year-old male from Karnataka and a 68-year-old female from Delhi, India were reported.

Figure 3.1 shows the distribution of confirmed COVID-19 cases in different states of India as of April 27, 2020. A total of 28,062 tested positive for COVID-19 in India as of April 27, 2020 with a total depth of 939 as of April 27, 2020.

According to the Indian Medical council of Research, 80% of Indian cases are found to be asymptomatic. The greatest challenge is tracking the asymptomatic cases since there is no other way than contact tracing. Recently by April 18, 186 new cases were tested positive with asymptomatic COVID-19. The initial study revealed by WHO says 80% of the cases were mild or asymptomatic 15% had severe respiratory syndrome and 5% were critical with the need for a ventilator.

3.2.1 INITIAL PREVENTIVE MEASURES WERE TAKEN BY THE GOVERNMENT OF INDIA AGAINST COVID-19

- MoHFM and the Ministry of Civil Aviation conducted screening tests in 21 airports of India. Initial screening was performed on

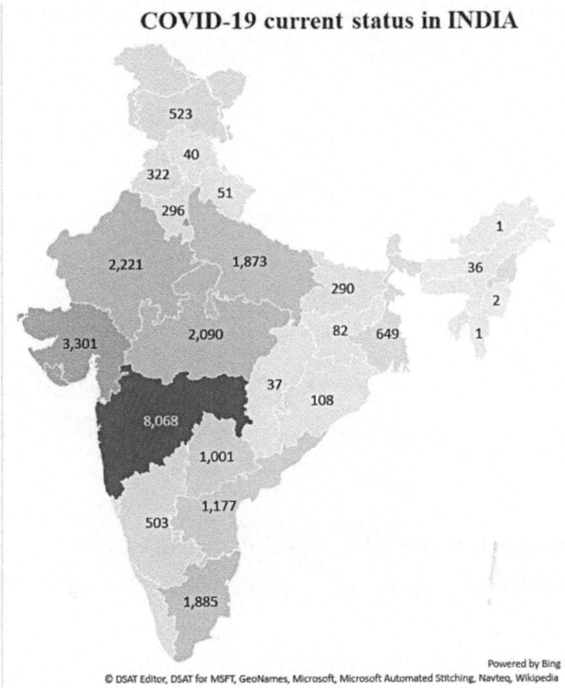

FIGURE 3.1 Distribution of laboratory confirmed cases of COVID-19 in India state wise as on April 27, 2020.

Source: Adapted from Ref. [18]

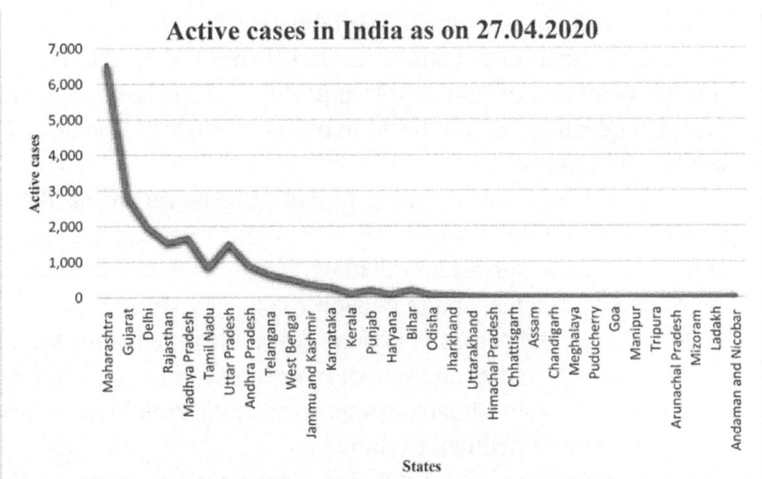

FIGURE 3.2 Active COVID-19 cases in Indian states as on April 27, 2020.

Source: Adapted from Ref. [18]

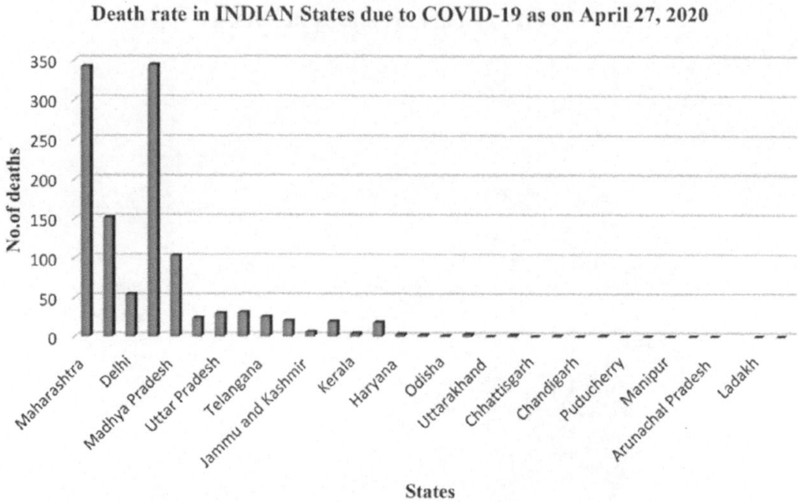

FIGURE 3.3 Total death in India due to COVID-19 as on April 27, 2020.
Source: Adapted from Ref. [18]

travelers from China who claimed to have symptoms like fever and cough.

- MoHFM has designated ICMR national Institute of Virology (NIV), Pune, as a nodal center for conducting a diagnosis of novel coronavirus and 11 more regional laboratories were trained to test the COVID-19 samples.
- A 24 × 7 control room was established by National Centre for Disease Control for the public to clarify any of their queries.
- The Government of India (GoI) and MoHFM has issued travel advisories requesting the public to avoid non-essential travel to China and other countries.
- Awareness was created among Indian citizens regarding the novel coronavirus. The symptoms and precautions to be taken to avoid the infection and its spread are dictated through social media, newspapers, advertisements, and television.
- A total lockdown across the country was informed from March 27, 2020, to date. Varsities and schools were closed in closed and the end semester exams and board exams for school kids were postponed and tentatively scheduled by June.
- Janata curfew and social distancing were issued as guidelines by the government to take control of the situation.

3.2.2 SYMPTOMS

Recently CDC has updated the symptoms of the corona virus and the symptoms may vary from person to person depending on their age, health, pre-existing condition.[9]

- Loss of smell and taste
- Intense chills
- Muscle pain
- Headache
- Pink eye
- Sore throat
- Fever
- Shortness of breath

With growing concern, it is observed that nCoV virus spreads not only through respiratory droplets or nasal but also gets transmitted through eye fluids causing conjunctivitis.

3.2.3 TREATMENT FOR COVID-19

At present, there is no proper treatment or vaccine available to treat the novel coronavirus. Researchers have strived hard and developed a number of COVID-19 vaccines all around the world. Some of the COVID-19 vaccines that are approved by WHO are Pfizer, AstraZeneca, Covishield, Janssen, Moderna, Sputnik, and so on.

3.3 LITERATURE REVIEW

In Rajinikanth et al.,[27] this proposed chapter aims to extract and evaluate the coronavirus disease (COVID-19) caused pneumonia infection in the lung using CT scans. It proposes an image-assisted system to extract COVID-19 infected sections from lung CT scans (coronal view). It includes the following steps: (i) threshold filter to extract the lung region by eliminating possible artifacts; (ii) image enhancement using harmony-search-optimization and Otsu thresholding; (iii) image segmentation to extract infected region(s); and (iv) region-of-interest (ROI) extraction (features) from the binary image to compute the level of severity. The features that are extracted from ROI are then employed to identify the pixel ratio between the lung and infection sections to identify infection level of severity. The proposed study is investigated in the MATLAB environment and the essential CTSI of COVID-19 is collected from the Radiopedia database. This work[22] implements a series of procedures, such as threshold filtering, multi-thresholding, segmentation, and area feature extraction to detect the COVID-19 pneumonia infection from the considered CTSI. After extracting the infected region with better accuracy, the infection stage is then computed using the ratio of pixels in the infection/lung section. This severity-level-measure (SLM) is then considered to short the existing patient's CTSI and the image that has a higher value of SLM will be sent to the pulmonologist for approval. The pulmonologist will also perform a visual check on the CTSI, and as per the findings, the appropriate treatment will be initiated.

In He et al.,[10] a publicly available dataset containing hundreds of CT scans positive for COVID-19 is built and sample-efficient deep learning methods are developed that can achieve high diagnostic accuracy of COVID-19 from CT scans even when the number of training CT images are limited. Specifically, it proposes a self-trans approach, which synergistically integrates contrastive self-supervised learning with transfer learning to learn powerful and unbiased feature representations for reducing the risk of overfitting. Extensive experiments demonstrate the superior performance of this proposed self-trans approach compared with several state-of-the-art baselines. This approach achieves an F1 of 0.85 and an AUC of 0.94 in diagnosing COVID-19 from CT scans, even though the number of trainings CTs is just a few hundred. It proposes a self-trans approach, which integrates contrastive self-supervision into the transfer learning process. Self-supervised learning is a learning paradigm that aims to capture the intrinsic patterns and properties of input data[29] (e.g., CT images) without using human-provided labels. The basic idea of SSL is to construct some auxiliary tasks solely based on the data itself without using human-annotated labels and force the network to learn meaningful representations by performing the auxiliary tasks well. Typical self-supervised learning approaches generally involve two aspects: constructing auxiliary tasks and defining loss functions. The auxiliary tasks are designed to encourage the model to learn meaningful representations of input data without utilizing human annotations. The loss functions are defined to measure the difference between a model's prediction and a fixed target, the similarities of sample pairs in a representation space (e.g., contrastive loss), or the difference between probability distributions (e.g., adversarial loss).

In Mishra et al.,[21] the positive images were collected from medRxiv and bioRxiv. These CT images are in different sizes corresponding to height (maximum—1853, average—491, and minimum—153), and width (maximum—1485, average—383, and minimum—124). Some sample COVID-19 positive and negative CT images were used. To prepare the final dataset for experiments, all the images have been converted into portable network graphics (.png) format to keep homogeneous characteristics. Further, both positive and negative class images were resized to $224 \times 224 \times 3$.

Recent advancements in the field of DL, especially in the medical imaging domain, indicate the potential usage of various deep CNN architectures. These baseline models include VGG16, InceptionV3, ResNet50, DenseNet121, and DenseNet201.[33] In this work, all of these baseline models' convolution parts are kept the same as the standard models, as proposed originally for the ImageNet challenge; however, the fully connected parts of the models are fixed as three fully connected layers (4096, 4096, and 1000), each with ReLU activation and finally a single-node prediction layer with Sigmoid activation function. Apart from these baseline models, a decision fusion-based approach is also considered in this work. The decision fusion model outperforms each of the individual models, achieving the highest average accuracy, AUC, and F1-Score of 0.8834, 0.8832, and 0.867, respectively. Among the individual models, DenseNet121 performs the best.

From the state-of-the-art experimental results, it is clear that deep CNN-based predictive models can be of very high value concerning COVID-19 identification from CT images. More specifically, the simple idea of decision fusion can improve the performance of deep CNN models quite drastically, achieving above 86% results to all the performance metrics under consideration. One more important observation is that the proposed approach has a very good reduction of false positive rate, which suggests its potential for use in real screening scenarios.

3.3.1 APPROACH—DEEP LEARNING MODEL TO DETECT COVID-19 USING CT SCAN

Deep learning comes under the umbrella of machine learning in artificial intelligence and has networks that are capable of unsupervised learning from data that is either unstructured or unlabeled.[4] Ever since there was news about the outbreak of COVID-19, there have consistently been more efforts to develop and hone deep learning methods for the task of performing screening of COVID-19 based on medical images including CT scans and X-ray images[5] of the chest as the virus is mainly known to affect the respiratory abilities of an infected person. Pre-processing techniques can help eliminate any kinds of unwanted noise and to further emphasize the aspects of the medical image thereby assisting with the recognition task, or even aid in the deep learning training phase. Deep learning is one of the most efficient techniques which can be utilized in medical science and technology. It is fast, it is efficient, and can be understood. It is a good method for diagnosis as well as prognosis of several illnesses and ailments with a good accuracy and precision rate. There are specifically trained models to classify the inputs given into different categories as intended by the programmers in question. In the medical field, they can be used for the detection of cardiovascular problems, tumors in the human body by using image analysis, diagnosing different types of cancer, and many other applications. It is used to differentiate the computerized tomography scan images of the patients infected with COVID-19 as positive or not. In the case of convolutional network models, the images taken as input are often resized to maintain their compatibility with the architecture's characteristic of the network. Currently, only a few large-sized datasets containing medical images related to COVID-19 are available for public use owing to increasing concerns about privacy and a blockade of information.

The automated diagnosis of COVID-19[26] from the computerized tomography scan pictures can be used by doctors and medical staff as a quick and efficient method for the screening of the virus. Early diagnosis of the disease is important to control this global pandemic.

3.4 MATERIALS AND METHODS

3.4.1 DATASET

There are many datasets to choose from but CT Scans have shown reliability and consistency rather than RT-PCR tests, X-ray images, or any other source of datasets. But, since this is a matter of privacy concerns, a publicly available dataset is hard to find which also has a large volume of images. However, the dataset obtained for experiments and testing is from Kaggle, an online platform where datasets are available for the public and also has a forum for enthusiasts to interact with each other and clear doubts and host discussions.

The current dataset contains 2482 images in which, 1252 are classified as COVID and the other 1230 are classified as NON-COVID. This is a relatively smaller dataset compared to others but is the largest publicly available dataset during the time of obtaining the dataset.

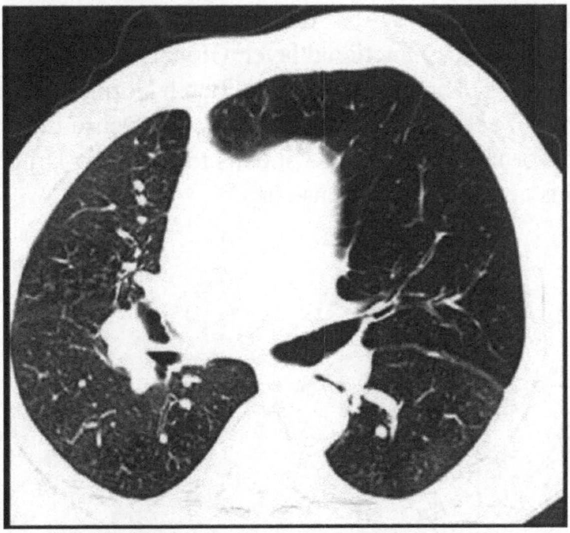

FIGURE 3.4 An image of a COVID-positive CT scan.

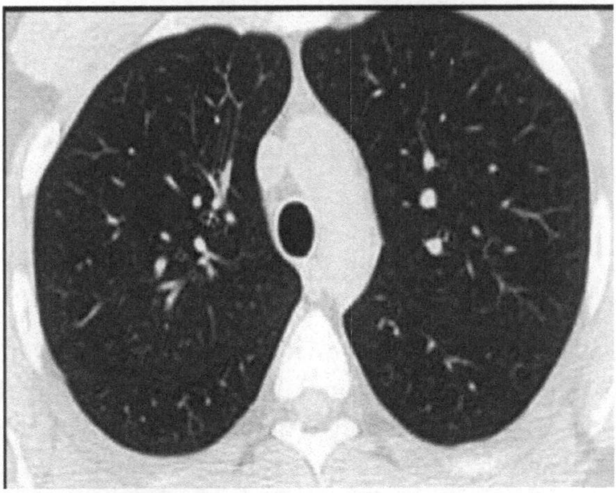

FIGURE 3.5 An image of COVID-negative CT scan.

3.4.2 PROPOSED ARCHITECTURE

The proposed architecture for this experiment is concluded on the finished result of the literature review. Most of the deep learning models in the referenced papers had an accuracy of not less than 90%. The only model that has been consistently crossed 90% was the DenseNet121 model. In this chapter, we use a form of DenseNet for our predictions. In this DenseNet model, there consists of an input layer for inputs to be defined. Then comes the

Conv2D layer which takes 64×64×64 input. The next layer is the DenseNet 121 layer which acts as a Functional layer. Global Average Pooling 2D layer is the following layer. The next layer is the Batch Normalization layer which is used to normalize the dataset. Then there is the Dense Layer followed by another Batch Normalization layer, which is followed by Dropout layer. The last layer of this network is the Dense layer.

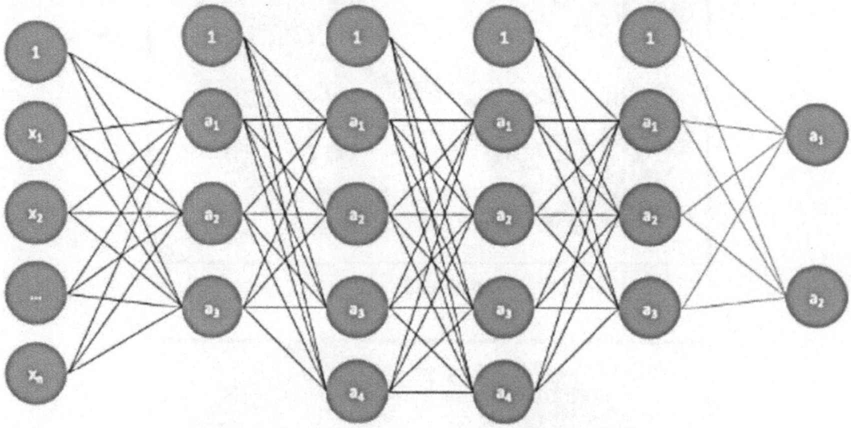

FIGURE 3.6 A sample image of convolutional neural network architecture.

Source: Reprinted from Jeremy Jordan. https://www.jeremyjordan.me/convolutional-neural-networks/

3.5 EXPERIMENT RESULTS

The experiments were conducted on Google Colab Notebook having an active GPU running in the background. The machine hosting this Notebook has Intel i7 Chip, 8 GB RAM, and Windows 10 OS. The total number of Epochs set is 150 and the image size is set to 64. The images in the dataset are modified with Image Data Generator and rotation ranges, width shift ranges, heights shift ranges, zoom ranges, etc., are added for the consistency and volubility of the dataset for training.

The optimizer is taken as Adam with the following parameters—lr=0.002, beta_1=0.9, beta_2=0.999, epsilon=0.1, decay=0.0.

The accuracy obtained at the end was 96.04% on the training set and 95.17% on the testing set for the DenseNet121 Model. This accuracy is one of its kind and is way better compared to other models proposed in other papers. An Accuracy and Loss Plot have been drawn after the experiment to check.

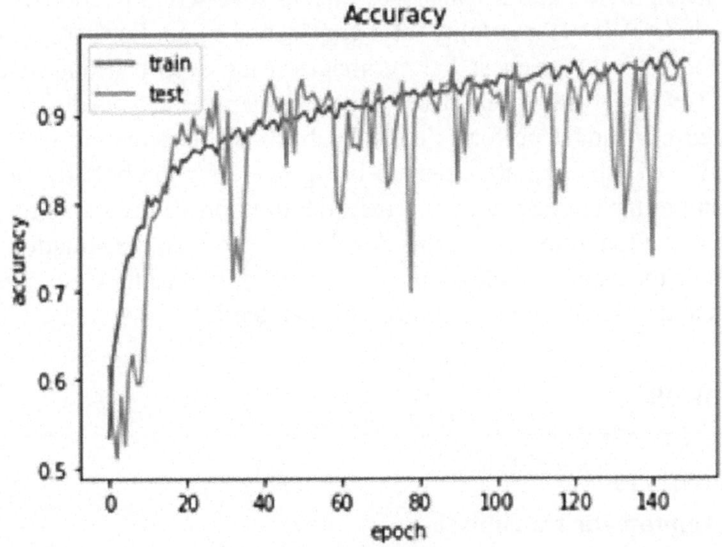

FIGURE 3.7 Accuracy vs Epochs.

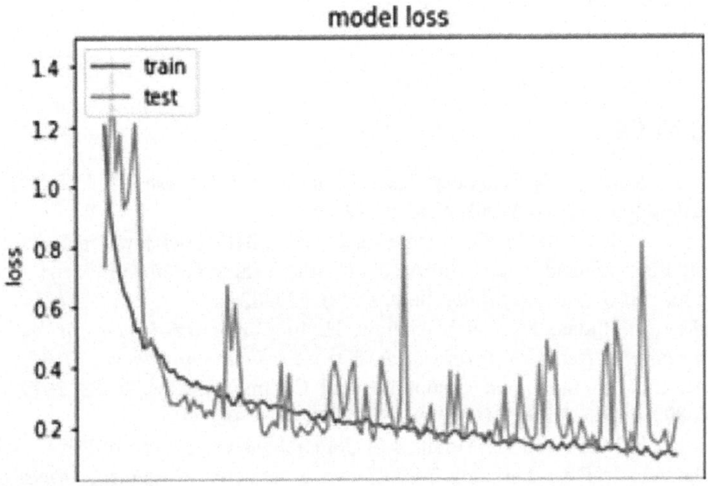

FIGURE 3.8 Model loss.

3.6 CONCLUSION

In this chapter, we presented the details of the COVID-19 outbreak with the facts of the world and more specific to India. We presented a detailed case study on the transmission of COVID-19 from person to person. The chapter also provides several analyses such as state-wise infection, infection rate graph, COVID-19 mortality ratio graphs, and so on. Further, we experimented on COVID-19 chest X-Ray images using deep learning models to detect COVID-19 infection. Based on the experimented results, it can be concluded our model performs considerably well in detecting COVID-19 infection. Nevertheless, attempts are being made for the betterment of the model and to try out new forms of research to improve the standards. Until new ways are found to control this deadly Coronavirus, the situation is still going to be the same. Vaccinations are a viable option as they are immunity boosters and must be given to control this pandemic.

KEYWORDS

- **COVID-19**
- **symptomatic transmission**
- **asymptotic transmission**
- **deep learning, chest X-ray**
- **prediction**

REFERENCES

1. Act, D. *Situation in Numbers*; Novel Coronavirus Disease (COVID-19) Situation Update Report-11, *11*(April), 2020, pp 12–15.
2. Ahmed, S.; Khaium, M. O.; Tazmeem, F. COVID-19 Lockdown in India Triggers a Rapid Rise in Suicides due to the Alcohol Withdrawal Symptoms: Evidence from Media Reports. *Int. J. Soc. Psychiatry* **2020,** *66* (8), 827–829.
3. Andrew, J.; Fiona, R.; Caleb Andrew, H. In *Comparative Study of Various Deep Convolutional Neural Networks in the Early Prediction of Cancer*, 2019 International Conference on Intelligent Computing and Control Systems, ICCS 2019, 2019, pp 884–890. https://doi.org/10.1109/ICCS45141.2019.9065445
4. Andrew, J.; Mathew, S. S.; Mohit, B. A Comprehensive Analysis of Privacy-Preserving Techniques in Deep Learning Based Disease Prediction Systems. *J. Phys. Conf. Ser.* **2019,** *1362* (1), 012070.
5. Andrew, J.; Mhatesh, T. S. R.; Sebastin, R. D.; Sagayam, K. M.; Eunice, J.; Pomplun, M.; Dang, H. Super-Resolution Reconstruction of Brain Magnetic Resonance Images via Lightweight Autoencoder. *Inform. Med. Unlocked* **2021,** *26,* 100713.
6. Andrew Onesimu, J.; Karthikeyan, J. An Efficient Privacy-Preserving Deep Learning Scheme for Medical Image Analysis. *J. Inform. Technol. Manag.* **2021,** *12* (Special Issue: The Importance of Human Computer Interaction: Challenges, Methods and Applications.), 50–67.

7. Anwar, T.; Zakir, S. In *Deep Learning Based Diagnosis of COVID-19 Using Chest CT-Scan Images*, Proceedings - 2020 23rd IEEE International Multi-Topic Conference, INMIC 2020, 2020. DOI: 10.1109/INMIC50486.2020.9318212
8. Backer, J. A.; Klinkenberg, D.; Wallinga, J. Incubation Period of 2019 Novel Coronavirus (2019- nCoV) Infections Among Travellers from Wuhan, China, 20 28 January 2020. *Eurosurveillance* **2020,** *25* (5). DOI: 10.2807/1560-7917.ES.2020.25.5.2000062
9. Gupta, N.; Agrawal, S.; Ish, P.; Mishra, S.; Gaind, R.; Usha, G.; Singh, B.; Sen, M. K.; group, *Safdarjung Hospital COVID 2019 Working. Clinical and Epidemiologic Profile of the Initial COVID-19 Patients at a Tertiary Care Centre in India. *Monaldi Arch. Chest Dis.* **2020,** *90* (1), 193–196.
10. He, X.; Yang, X.; Zhang, S.; Zhao, J.; Zhang, Y.; Xing, E.; Xie, P. Sample-Efficient Deep Learning for COVID-19 Diagnosis Based on CT Scans. *IEEE Trans. Med. Imag.* **2020,** *XX* (Xx). DOI: 10.1101/2020.04.13.20063941
11. Joy, R. C.; Sagayam, K. M.; Rajan, A. A.; Andrew, J.; Ho, C. C. Analyzing Human Emotion from Facial Expression Based on Eigen Face Approach. *Int. J. Sci. Technol. Res.* **2020,** *9* (1), 3247–3250.
12. Karim, M. R.; Dohmen, T.; Cochez, M.; Beyan, O.; Rebholz-Schuhmann, D.; Decker, S. In *Deep COVID Explainer: Explainable COVID-19 Diagnosis from Chest X-ray Images*, Proceedings - 2020 IEEE International Conference on Bioinformatics and Biomedicine, BIBM, 2020, 1034–1037. DOI: 10.1109/BIBM49941.2020.9313304
13. Lancet, T. India Under COVID-19 Lockdown. *Lancet (London, England)*, **2020,** *395* (10233), 1315.
14. Li, Q.; Guan, X.; Wu, P.; Wang, X.; Zhou, L.; Tong, Y.; Ren, R.; Leung, K. S. M.; Lau, E. H. Y.; Wong, J. Y.; Xing, X.; Xiang, N.; Wu, Y.; Li, C.; Chen, Q.; Li, D.; Liu, T.; Zhao, J.; Liu, M.; … Feng, Z. Early Transmission Dynamics in Wuhan, China, of Novel Coronavirus–Infected Pneumonia. *N. Engl. J. Med.* **2020,** *382* (13), 1199–1207.
15. Liu, D. X.; Liang, J. Q.; Fung, T. S. Human Coronavirus-229E, -OC43, -NL63, and -HKU1 (Coronaviridae). *Encyclopedia Virol.* **2021,** 428–440. DOI: 10.1016/B978-0-12-809633-8.21501-X
16. Liu, Y.; Yan, L. M.; Wan, L.; Xiang, T. X.; Le, A.; Liu, J. M.; Peiris, M.; Poon, L. L. M.; Zhang, W. Viral Dynamics in Mild and Severe Cases of COVID-19. *Lancet Infect. Dis.* **2020,** *20* (6), 656–657.
17. Livingston, E.; Bucher, K.; Rekito, A. Coronavirus Disease 2019 and Influenza 2019-2020. *JAMA* **2020,** *323* (12), 1122.
18. Macedonia, N.; Prime, T.; Secretary, C.; Africa, S. *Global Updates India Situation Situation in Kerala*. **2020,** *2020* (February), 26–28.
19. Mertens, P.; De Vos, N.; Martiny, D.; Jassoy, C.; Mirazimi, A.; Cuypers, L.; Van den Wijngaert, S.; Monteil, V.; Melin, P.; Stoffels, K.; Yin, N.; Mileto, D.; Delaunoy, S.; Magein, H.; Lagrou, K.; Bouzet, J.; Serrano, G.; Wautier, M.; Leclipteux, T.; … Vanderlinden, J. Development and Potential Usefulness of the COVID-19 Ag Respi-Strip Diagnostic Assay in a Pandemic Context. *Front. Med. (Lausanne)* **2020,** *7*, 225.
20. Mhathesh, T. S. R.; Andrew, J.; Martin Sagayam, K.; Henesey, L. A 3D Convolutional Neural Network for Bacterial Image Classification. In *Advances in Intelligent Systems and Computing,* Vol. 1167; Springer, 2021; pp 419–431.
21. Mishra, A. K.; Das, S. K.; Roy, P.; Bandyopadhyay, S. Identifying COVID19 from Chest CT Images: A Deep Convolutional Neural Networks Based Approach. *J. Healthc. Eng.* **2020,** *2020,* 8843664.

22. Motamed, S.; Rogalla, P.; Khalvati, F. RANDGAN: Randomized Generative Adversarial Network for Detection of COVID-19 in Chest X-ray. *Sci. Rep.* **2021,** *11* (1), 1–10.
23. Onesimu, J. A.; J.; K.; Viswas, D. S. J.; Sebastian, R. D. Security and Privacy Challenges of Deep Learning. In *Research Anthology on Privatizing and Securing Data*; 2021; pp 1258–1280. DOI: 10.4018/978-1-7998-8954-0.ch059
24. Onesimu, J. A.; Karthikeyan, J.; Sei, Y. An Efficient Clustering-Based Anonymization Scheme for Privacy-Preserving Data Collection in IoT Based Healthcare Services. *Peer-to-Peer Netw. Appl.* **2021,** *14* (3), 1629–1649.
25. Prime, T. *India Situation Situation in Kerala*. **2020,** *February,* 1–4.
26. Punn, N. S.; Agarwal, S. Automated Diagnosis of COVID-19 with Limited Posteroanterior Chest X-Ray Images Using Fine-Tuned Deep Neural Networks. *Appl. Intell (Dordr)*. **2021,** *51* (5), 2689–2702.
27. Rajinikanth, V.; Dey, N.; Raj, A. N. J.; Hassanien, A. E.; Santosh, K. C.; Raja, N. S. M. *Harmony-Search and Otsu based System for Coronavirus Disease (COVID-19) Detection using Lung CT Scan Images* [Online], 2020. https://arxiv.org/abs/2004.03431v1
28. Rothe, C.; Schunk, M.; Sothmann, P.; Bretzel, G.; Froeschl, G.; Wallrauch, C.; Zimmer, T.; Thiel, V.; Janke, C.; Guggemos, W.; Seilmaier, M.; Drosten, C.; Vollmar, P.; Zwirglmaier, K.; Zange, S.; Wölfel, R.; Hoelscher, M. Transmission of 2019-nCoV Infection from an Asymptomatic Contact in Germany. *N. Engl. J. Med.* **2020,** *382* (10), 970–971.
29. Ter-Sarkisov, A. Lightweight Model for the Prediction of COVID-19 through the Detection and Segmentation of Lesions in Chest CT Scans. *MedRxiv*, **2020,** *2* (1), 1–15.
30. Ter-Sarkisov, A. One Shot Model for the Prediction of COVID-19 and Lesions Segmentation in Chest CT Scans Through the Affinity Among Lesion Mask Features. *MedRxiv*, **2021,** 1–14.
31. USAID. *India: Situation Analysis*. **2020,** *2* (February), 1–4.
32. Van Erum, J.; Van Dam, D.; De Deyn, P. P. Alzheimer's Disease: Neurotransmitters of the Sleep-Wake Cycle. *Neurosci. Biobehav. Rev.* **2019,** *105*, 72–80.
33. Wang, L.; Lin, Z. Q.; Wong, A. COVID-Net: A Tailored Deep Convolutional Neural Network Design for Detection of COVID-19 Cases from Chest X-ray Images. *Sci. Rep.* **2020,** *10* (1), 1–12.
34. Weiss, S. R.; Navas-Martin, S. Coronavirus Pathogenesis and the Emerging Pathogen Severe Acute Respiratory Syndrome Coronavirus. *Microbiol. Mol. Biol. Rev.* **2005,** *69* (4), 635–664.
35. *What you Need to Know About Coronavirus Disease 2019 (COVID-19)*. **2020,** *2019,* 314937.
36. Woo, P. C. Y.; Lau, S. K. P.; Chu, C.-M.; Chan, K.-H.; Tsoi, H.-W.; Huang, Y.; Wong, B. H. L.; Poon, R. W. S.; Cai, J. J.; Luk, W.-K.; Poon, L. L. M.; Wong, S. S. Y.; Guan, Y.; Peiris, J. S. M.; Yuen, K.-Y. Characterization and Complete Genome Sequence of a Novel Coronavirus, Coronavirus HKU1, from Patients with Pneumonia. *J. Virol.* **2005,** *79* (2), 884–895.

CHAPTER 4

Shortest Path Problem Using Warshall–Floyd Algorithm in Spherical Fuzzy Environment: An Outlook in Government-Controlled Transportation System in COVID-19

ANKAN BHAUMIKA

Mathematics Department, Dhamcha Chhagulia Siddheswari High School, Paschim Medinipur, West Bengal, India

ABSTRACT

In this chapter, we consider the most interesting areas of network analysis in Operations Research—Shortest Path Problem. Here, we assume the solution methodology of the Warshall–Floyd algorithm with modification. Basically, in the Warshall–Floyd algorithm, when the arcs are directional, no passage is permitted in the reverse direction. But, in real-life problems, reverse direction is possible, and in such cases, we consider the cost amount corresponding to reverse passage as a big one, say, 'M'. This modification is included in this chapter. Now, the whole world is under the shadow of COVID-19. We must chalk out all possible ways so that we can be free from the grip of this pandemic situation. For this reason, governments and nongovernment organizations are trying their best in both hemispheres. Besides the uses of mask-soap-sanitizer-maintain distances, lockdown is another weapon against the chain of COVID-19 viruses. In the time of lockdown, the government totally stopped the public transportation system. Recently, in many areas, transportation is going to be normal. But, in India, mainly in West Bengal, local trains are not running as usual as prelockdown times. That's why general people

Global Digital Transformation and the Covid-19 Pandemic.
Rajdeep Chowdhury, K. Martin Sagayam, Neyara Radwan, Ahmad A. Elngar (Eds.)

are using bus services, hiring autos, etc. This government-made situation is spreading COVID-19 viruses, which are established mathematically in this chapter. In this chapter, we consider the fare costs of distances as variables that are changed with some uncertain characteristics. This type of uncertainty is assumed here as interval-valued linguistic spherical fuzzy numbers, and we get some results. The results of our proposed approach are compared with other studies as a conclusive part of this chapter.

4.1 INTRODUCTION

In this chapter, we discuss an important subtopic of the topic of network analysis called the shortest path problem. A network is a systematic communication or path constituting lines or channels by which information is transferred. Some examples of networks around us are public transportation systems, oil-gas-water pipeline transportation, transportation of goods to different markets, aviation transportation, communication on the internet, mobiles or satellites, transmission of blood through veins, movement of traffic on roads, etc. Normally, a network system starts from the source (supply stations, which may be production storage hubs, companies, countries, etc.) and ends with the sink (demanding stations, which may be customers, retailers, local markets, etc.). Most of the real-world problems can be viewed as models under network analysis. These models are broadly used, in actuality, for some reasons. Firstly, networks are incredible representation apparatuses and straightforward, as issues can be introduced pictorially over networks. Secondly, network models have a few computationally effective and simple-to-carry-out calculations that exploit the exceptional construction of organizations, hence giving a vital benefit to specialists to deal with huge, genuine combinatorial issues. Moreover, network problems show up as subproblems when numerical programming methods like deterioration, row-column generation or alteration, etc. are used to tackle complex numerical models. This is a key benefit since utilizing productive and efficient network algorithmic calculations brings about quicker convergence when the above-stated methods are used. For instance, shortest path problems productively show up in a huge number in different cases in transportation, correspondence, logistics, supply chain management, internet directing, atomic science, physical science, humanism, etc.

Nowadays, the whole world is under the shadow of COVID-19. In the most recent 3 years, we have been under very strict restrictions on various demonstrated and proven scientific strategies as well as different trial-and-error procedures. Sanitizing and washing hands, keeping good distances from others, and trying not to be a part of any gathering are some of the proven scientific strategies to combat COVID-19. On the other hand, to stop the spread of infections, trains and public vehicles are not running consistently because of managerial or administrative choices in various parts of various nations. In West Bengal, India, we are under the trial-and-error method of such government decisions. In this chapter, we have established that this sort of choice of government tends to cause imperfections in the battle against COVID-19 utilizing numerical calculations. Though our mathematical studies have some advantages and disadvantages, this can show another way of reevaluating the halting of public transportation frameworks, mostly train services. For this real-life problem, in this chapter we have considered the spherical fuzzy set in its extended version, say, the interval-valued linguistic spherical fuzzy set (IVLSFS). We have considered a new ranking approach of interval-valued linguistic spherical fuzzy numbers (IVLSFNs) to tackle the corresponding mathematical problem easily. In this chapter, train-bus route problems are considered through some modification of the Warshall–Floyd problem under the IVLSFS concept.

4.1.1 NOTATIONS AND ASSUMPTIONS

$\breve{F}$: A fuzzy set in the universe of discourse χ;
$\hat{F}$: An intuitionistic fuzzy set;
$\grave{A}$: A spherical fuzzy set;
$\tilde{a}=\left[\tilde{a}^l,\tilde{a}^u\right]$: An interval fuzzy number;
$\acute{A}$: An interval-valued spherical fuzzy set;
$\acute{A}$: An IVLSFS.

4.1.2 SHORTEST PATH PROBLEM

Shortest path problems entail a common network design through which decision-makers wish to get the shortest route among different nodes satisfying minimum time, cost, or maximum reliability and efficiency. In the shortest path problem, minimum cost flows from the origin and ships units of flow to every other node on the network, considering the total traversed distance to be minimum. A shortest path problem can be considered a whole problem or can be divided into subproblems of a complex system. In a transportation problem, where we have to travel the shortest distance between two places, or in a telecommunication problem, when we convey messages between nodes quickly, we generally use shortest path problems. Transportation of oil and gas from refineries to fuel stations or from dealers to customers and water distribution systems from water reservoirs to consumers are very interesting shortest path problems. Shortest path problems are considered in the replacement of items, project management, cash flow management in shopping malls, work planning in industry, inventory and production planning in business management, and even in genome sequencing. Shortest path problems can be assumed as subproblems in traffic management for different network issues.

Shortest path problems can be classified as undirected, directed, or mixed. Several algorithms are available to solve these types of shortest path problems. These algorithms are the Warshall–Floyd algorithm, Seidel's algorithm, the Bellman–Ford algorithm, Dijkstra's algorithm, etc. Among these, in this chapter, we have considered the Warshall–Floyd algorithm.

4.1.3 COVID-19 AND TRANSPORTATION SYSTEMS

Today's world is a communicative world, and the main physical source of public communication is the public transportation system. The main part of this system consists of train and bus services. The impacts of a major pandemic such as COVID-19 on public transit systems in different countries are tremendous and cause the decline of both consumers and transportation servers (Liu et al., 2020; Bliss, 2020; Krishnakumari and Cats, 2020). The road transport, that is, the train and bus transportation sector, is crucial for social and monetary advancement and ensures versatility across nations. Road cargo transport services guarantee the stock of fundamental products, including food and clinical supplies, as well as a huge range of materials on which organizations and purchasers depend. The road transportation system not only makes a significant commitment to monetary development and occupation creation, but also forestalls deficiencies in fundamental merchandise. Road transportation services in different nations have been announced as fundamental services. To control the spread of COVID-19, numerous nations have put limitations on domestic transit and/or closed border crossings for road transportation services. The subsequent slump in exchange and inventory network disturbances will have an effect on all nations and may additionally increase the weakness of landlocked agricultural nations, mainly on women, the elderly, and informal workers. A few variables add to unveiling transportation stations and vehicle conditions high danger for the COVID-19 virus:

- Individuals are restricted to closed spaces in trains and buses. So, infection hazards increase with the degree of traveler occupancy in vehicles and stations.

- There may be scant access control to recognize travelers or laborers who might be COVID-19 positive.
- The presence of numerous surfaces, like seats, handrails, entryways, and ticket machines, can effectively move microbes.

For these reasons, the government and local administrations are unwilling to support public transportation systems. In West Bengal, India, government authorities also distribute a circular (West Bengal Govt. Order, 2020) to stop the public transportation system, mainly train services.

In this chapter, we discuss the IVLSFS-based environment and consider a modified Warshall–Floyd algorithmic approach to IVLSFS in the shortest path problem. The study of the public transportation system is considered here mathematically with uncertainty in fuzzy variables with linguistic form.

The main highlights of this chapter can be characterized as:

- This chapter describes the COVID-19 transportation problem under IVLSFSs.
- The COVID-19 transportation problem is solved in this chapter using the Warshall–Floyd algorithm with modifications.
- A new de-spherifuzzification approach is used in this chapter.
- The approach to solving the real-life problem shows the flaws of the government's decision on public transportation in the COVID-19 situation.

The residues of this chapter are as follows: Section 4.2 elaborately describes the previous research works about fuzzy sets, interval numbers, linguistic term sets, spherical fuzzy sets, etc. Section 4.3 describes the preliminaries, that is, the definitions, examples, properties, etc. of fuzzy sets, intuitionistic fuzzy sets, interval numbers, interval-valued spherical fuzzy sets, and linguistic term sets. Section 4.4 describes the IVLSFS. A new ranking approach for IVLSFS is defined in Section 4.5. The algorithm of the Warshall–Floyd-based shortest route finding problem is briefly studied in Section 4.6. Section 4.7 is engaged with a case study on COVID-19 and transportation systems in West Bengal, India, with the obtained results. The conclusion, advantages, and disadvantages with future research scopes are depicted in Section 4.8.

4.2 LITERATURE REVIEW

The fuzzy sets (Zadeh, 1965) and the concept of fuzziness of different aspects play important roles when we consider real-life problems. Obviously, consideration of optimal choices is based on the fuzziness of the considered data. In real-life problems, variables are expressed in linguistic modes or terms and are more acceptable to decision-makers and society than numerical data. These variables, termed linguistic variables (Zadeh, 1975), strengthen their relevancy under real-life uncertainties in various fields after the proposal of linguistic information in fuzzy set theory. The two-dimensional logical representations of fuzzy sets sometimes become insufficient to tackle all the uncertainties prevailing in the system. In different cases, a single value is inadequate to provide complete information about the degree of kinship of an element of the considered set. Torra (2010) coined the term hesitant fuzzy set as another accretion of fuzzy set where every member can be defined with a set of membership degrees, each. A hesitant fuzzy set is considered when decision-makers feel a hesitant environment, like disaccord or discrepancy situations. Xia and Xu (2011) described some operations on hesitant fuzzy sets. Chen and Xu (2014) defined the properties of interval-valued hesitant fuzzy sets. Chen et al. (2013) considered interval-valued hesitant preference relations with related applications into group decision-making problems. Herrera and Herrera-Viedma (2000) made an analysis on stepwise solving decision problems under linguistic information. A fuzzy linguistic methodology was proposed by Herrera et al. (2008) to deal with unbalanced linguistic term sets. Herrera et al. (1995) considered a sequential selection process in group decision-making with linguistic assessment. Jiang and Wei (2014) depicted some Bonferroni mean operators with 2-tuple linguistic information and applied these operators to multiple attribute decision-making. Kacprzyk and Zadrozny (2005) considered linguistic information in knowledge discovery tools. Several articles (Martinez et al., 2005; Xu, 2005, 2004) have been published on fuzzy linguistic information and hesitant fuzzy linguistic information. Yager (1997) considered multiple objective decision-making problems using fuzzy sets. Yu et al. (2013) introduced a hesitant fuzzy environment into personnel evaluation. Rodriguez et al. (2012) elaborately studied hesitant fuzzy linguistic term sets for decision-making. Rodriguez and Martinez (2013) analyzed symbolic linguistic computing models in decision-making. Özkan et al. (2019) successfully introduced a hesitant fuzzy linguistic MCDM approach to a real-life problem. Numerous articles (Bhaumik and Roy, 2019; Bhaumik et al., 2020a, b, 2021; Roy and Bhaumik, 2018) have been published using fuzzy, intuitionistic fuzzy, hesitant fuzzy, and linguistic information.

Several articles (Ashraf and Abdullah, 2019; Ashraf et al., 2019a; Jin et al., 2019; Liu et al., 2020a, b; Rafiq et al., 2019) have been published on decision-making problems using spherical fuzzy sets and their different extended versions. A spherical fuzzy set has been used to solve several real-life problems. Ashraf et al. (2019b) determined the environmental development factors of children using a spherical fuzzy set. Boltürk (2019) considered AS/RS technology selection using a spherical fuzzy concept. Gündoğdu and Kahraman (2019b) applied a spherical fuzzy hierarchy process in renewable energy applications. Spherical fuzzy sets and concepts have been used from industrial robot selection (Gündoğdu and Kahraman, 2019c) to warehouse site selection (Gündoğdu and Kahraman, 2019d), from public evaluation of cycling (Liu et al., 2020c) to medical diagnosis (Mahmood et al., 2019), etc.

Several articles (Liu et al., 2020; Bliss, 2020; Krishnakumari and Cats, 2020; Tirachini and Cats, 2020) have been written on public transportation systems based on the COVID-19 pandemic situation, but strategies of public transportation in COVID-19 by administration are criticized here using the Warshall–Floyd-based shortest path problem under the IVLSFS concept, which is new in the literature.

4.3 MATHEMATICAL PRELIMINARIES

In this section, we consider the basics of fuzzy sets, intuitionistic fuzzy sets, spherical fuzzy sets, interval numbers, interval-valued spherical fuzzy sets, and linguistic term sets with examples and properties.

Definition 4.3.1. Let χ be a universe of discourse. A fuzzy set $\breve{F}\,(\in\chi)$ is characterized by a membership function $\mu_{\breve{F}}: \chi \rightarrow [0,1]$ and is defined by:

$$\breve{F}=\left\{\left(x,\mu_{\breve{F}}\left(x\right)\right):\mu_{\breve{F}}\left(x\right)\in\left[0,1\right],x\in\chi\right\}.$$

Obviously, the membership degrees $\mu_{\breve{F}}$: (x) of $\breve{F}$ are crisp numbers.

Definition 4.3.2. (Atanassov, 1986) An intuitionistic fuzzy set (IFS), $\hat{F}$ in χ is given by,

$$\hat{F}=\left\{\left\langle x,\mu_{\hat{F}}\left(x\right),\gamma_{\hat{F}}\left(x\right)\right\rangle:x\in\chi\right\},$$

where both $\mu_{\hat{F}}$ and $\gamma_{\hat{F}}$ maps the elements of χ to [0, 1], with $0 \le \mu_{\hat{F}}(x) + \gamma_{\hat{F}}(x) \le 1$, $\forall x \in \chi$. For each, $x, \mu_{\hat{F}}(x)$ and $\gamma_{\hat{F}}(x)$ represent the degree of membership and the degree of nonmembership, respectively. Again $\pi_{\hat{F}}(x) = 1 - \mu_{\hat{F}}(x) - \gamma_{\hat{F}}(x)$ is called the "*degree of hesitation*" of the element x in the set $\hat{F}$. If $\pi_{\hat{F}}(x) = 0, \forall x \in \chi$, then the IFS represents a fuzzy set.

Sometimes, FS or IFS are not used due to the complexity appearing in the assumed problem. The spherical fuzzy set (SFS) fills the gap where hesitation characteristics occur due to the concerned problem.

Definition 4.3.3. (Ashraf and Abdullah, 2019) A SFS $\grave{A}$ in the universe of discourse χ is defined as:

$\grave{A}=\left\{\left\langle x,\left(\rho_{\grave{A}}(x),\phi_{\grave{A}}(x),\psi_{\grave{A}}(x)\right)\right\rangle \mid x\in\chi\right\}$, where membership, nonmembership, and hesitancy degrees are described as: $\rho_{\grave{A}}(x):\chi\to[0,1]$, $\phi_{\grave{A}}(x):\chi\to[0,1]$ and $\psi_{\grave{A}}(x):\chi\to[0,1]$ with $0\le\left(\rho_{\grave{A}}(x)\right)^2+\left(\phi_{\grave{A}}(x)\right)^2+\left(\psi_{\grave{A}}(x)\right)^2\le 1, \forall x\in\chi$.

Here, $\xi=\sqrt{1-(\rho^2+\phi^2+\psi^2)}$ is called as a refusal degree to the corresponding spherical fuzzy number.

In the next part, some preliminaries about interval numbers, the expected value of an interval number, and interval fuzzy numbers are discussed.

Definition 4.3.4. Let $a=\left[a^l,a^u\right]=\left\{x:a^l\le x\le a^u\right\}$, then a is called an interval number. If $a^l\ge 0$, then a is called positive interval number.

Consider two positive interval numbers $a=\left[a^l,a^u\right]$ and $b=\left[b^l,b^u\right]$ with $\lambda\in[0,1]$ Then,

(i) $a+b=[a^l+b^l,a^u+b^u]$;

(ii) $a^\lambda=\left(\left[a^l,a^u\right]\right)^\lambda=\left[\left(a^l\right)^\lambda,\left(a^u\right)^\lambda\right]$;

(iii) $\lambda a=\lambda\left[a^l,a^u\right]=\left[\lambda a^l,\lambda a^u\right]$;

(iv) $a.b=\left[a^l,a^u\right].\left[b^l,b^u\right]=\left[a^l.b^l,a^u.b^u\right]$;

(v) $a\cup b=\left[\max\left\{a^l,b^l\right\},\max\left\{a^u,b^u\right\}\right]$;

(vi) $a\cap b=\left[\min\left\{a^l,b^l\right\},\min\left\{a^u,b^u\right\}\right]$.

Definition 4.3.5. Assume $a=\left[a^l,a^u\right]=\left\{x:0\le a^l\le x\le a^u\right\}$ an interval number, then the expected value of a is $E(a)=(1/2)\left(a^l+a^u\right)$.

Definition 4.3.6. Assume $\tilde{a}=\left[\tilde{a}^l,\tilde{a}^u\right]=$ $\{x:\tilde{a}^l\le x\le\tilde{a}^u,\tilde{a}^l=\inf\tilde{a},\tilde{a}^u$ sup $=\sup\tilde{a}\}$ as an interval fuzzy number, then the expected value of $\tilde{a}$ is defined as $E(\tilde{a})=(1/2)\left(\tilde{a}^l+\tilde{a}^u\right)$.

Definition 4.3.7. Let $\tilde{a}_1 = \left[\tilde{a}_1^l, \tilde{a}_1^u\right]$ and $\tilde{a}_2 = \left[\tilde{a}_2^l, \tilde{a}_2^u\right]$ be two interval fuzzy numbers with $l(\tilde{a}_1) = \tilde{a}_1^u - \tilde{a}_1^l$, $l(\tilde{a}_2) = \tilde{a}_2^u - \tilde{a}_2^l$; then the degree of possibility of $\tilde{a}_1 \geq \tilde{a}_2$ is defined as follows (eq 4.1):

$$Pos\left(\tilde{a}_1 \geq \tilde{a}_2\right) = \max\left\{1 - \max\left\{\frac{\tilde{a}_2^u - \tilde{a}_1^l}{l\left(\tilde{a}_1\right) + l\left(\tilde{a}_2\right)}, 0\right\}, 0\right\}. \tag{4.1}$$

Similarly, the degree of possibility of $\tilde{a}_2 \geq \tilde{a}_1$ is defined as (eq 4.2):

$$Pos\left(\tilde{a}_2 \geq \tilde{a}_1\right) = \max\left\{1 - \max\left\{\frac{\tilde{a}_1^u - \tilde{a}_2^l}{l\left(\tilde{a}_1\right) + l\left(\tilde{a}_2\right)}, 0\right\}, 0\right\}. \tag{4.2}$$

From Definition 4.3.7, we easily obtain the following results:

(i) $0 \leq Pos\left(\tilde{a}_1 \geq \tilde{a}_2\right) \leq 1$;

(ii) $0 \leq Pos\left(\tilde{a}_2 \geq \tilde{a}_1\right) \leq 1$;

(iii) $Pos\left(\tilde{a}_1 \geq \tilde{a}_2\right) + Pos\left(\tilde{a}_2 \geq \tilde{a}_1\right) = 1$; and specially,

(iv) $Pos\left(\tilde{a}_1 \geq \tilde{a}_1\right) = Pos\left(\tilde{a}_2 \geq \tilde{a}_2\right) = 0.5$.

Definition 4.3.8. Interval-Valued Spherical Fuzzy Set (Kahraman and Gündoğdu, 2021): An interval-valued spherical fuzzy set (IVSFS) A'' in the universe of discourse χ is defined as:

$$A'' = \left\{\left\langle x, \left(\left[\rho_{A''}^l(x), \rho_{A''}^u(x)\right], \left[\phi_{A''}^l(x), \phi_{A''}^u(x)\right], \left[\psi_{A''}^l(x), \psi_{A''}^u(x)\right]\right)\right\rangle \mid x \in \chi\right\}$$

where $\left[\rho_{A''}^l(x), \rho_{A''}^u(x)\right] \subseteq [0,1], \left[\phi_{A''}^l(x), \phi_{A''}^u(x)\right] \subseteq [0,1], \left[\psi_{A''}^l(x), \psi_{A''}^u(x)\right] \subseteq [0,1]$

and $0 \leq \left(s_{\rho_{A''}^u(x)}\right)^2 + \left(s_{\phi_{A''}^u(x)}\right)^2 + \left(s_{\psi_{A''}^u(x)}\right)^2 \leq 1, \forall x \in \chi$.

Variables, when expressed by languages, words, or sentences, are called linguistic variables. These variables enhance their applicability to real-life uncertainties in various fields. Zadeh (1975) proposed the linguistic term set.

Definition 4.3.9. Linguistic Term Set (Zadeh, 1975): Assume,

(a) *H*, the name of the variable;
(b) *T* (*H*), the term set of *H*, that is, the set of its linguistic values;
(c) *U*, a universe of discourse;
(d) *G*, the way by which the terms of *T* (*H*) are generated;

(e) *M*, a semantic rule for associating each linguistic value *X* with its meaning, that is, *M* (*X*), a fuzzy subset of *U*.

Then, a linguistic variable is characterized by a quintuple (*H*, *T* (*H*), *U*, *G*, *M*). A linguistic variable is linguistically described by its semantics.

Several expressions of linguistic descriptors and the corresponding semantics exist in literature (Herrera and Herrera-Viedma, 2000; Herrera et al., 2008; Rodriguez et al., 2012; Rodriguez and Martinez, 2013). Among these, seven scales of linguistic term-based semantics are frequently used, given as:

S_1 = {s0=nothing, s1=very low, s2=low, s3=medium, s4=high, s5=very high, s6=perfect} or, S2 = {s0=very poor, s1=poor, s2=slightly poor, s3=fair, s4=slightly good, s5= good, s6=very good}.

Property 4.3.10. The properties of linguistic term set $S=\{s_i : i= 1,2,...,t\}$ are:

(i) The set is ordered, that is, $s_i > s_j$, if $i > j$;

(ii) A negation operator, that is, $\neg(s_i)=s_j\,(\text{for } i+j=t+1)$ exists;

(iii) A maximizing operator, that is, $\max(s_i,s_j)=s_j(\text{if } s_j \geq s_i)$ exists;

(iv) A minimizing operator, that is, $\min(s_i,s_j)=s_j(\text{if } s_j \leq s_i)$ exists.

This discrete term set can be converted into a continuous term set as follows:

$\bar{S}=\{s_i : s_1 \leq s_i \leq s_q, i\in[1,q]\}$, where *q* is a sufficiently large positive number. s_i is called the original linguistic term if $s_i \in S$, otherwise, the virtual linguistic term.

Property 4.3.11. Assume s_γ and s_δ be two linguistic variables; $s_\gamma, s_\delta \in \bar{S}; \lambda, \kappa \in [0,1]$. The operational laws are defined as (Xu, 2004):

(i) $s_\gamma \oplus s_\delta = s_{\gamma+\delta}$;

(ii) $s_\gamma \otimes s_\delta = s_{\gamma\delta}$;

(iii) $\lambda s_\gamma = s_{\lambda\gamma}$;

(iv) $(\lambda+\kappa)s_\gamma = \lambda s_\gamma \oplus \kappa s_\gamma$;

(v) $(s_\gamma)^\lambda = s_{\gamma\lambda}$

Human decisions and perceptions consistently stream in uncertain environments like spherical fuzzy, and essentially, these environments are supported by linguistic characters of reactions having fuzziness sense. In the following, we exhibit the interval-valued linguistic term set-based spherical fuzzy set.

4.4 INTERVAL-VALUED LINGUISTIC SPHERICAL FUZZY SET (IVLSFS)

We expressed here the hesitant interval-valued fuzzy linguistic numbers through definition.

Definition 4.4.1. A IVLSFS $\acute{A}$ in the universe of discourse χ is defined as:

$$\acute{A}=\left\{\left\langle x,\left(\left[s_{\rho_{\acute{A}}^{l}}(x),s_{\rho_{\acute{A}}^{u}}(x)\right],\left[s_{\phi_{\acute{A}}^{l}}(x),s_{\phi_{\acute{A}}^{u}}(x)\right],\left[s_{\psi_{\acute{A}}^{l}}(x),s_{\psi_{\acute{A}}^{u}(x)}\right]\right)\right\rangle \mid x\in\chi\right\}$$

where $\left[s_{\rho_{\acute{A}}^{l}}(x),s_{\rho_{\acute{A}}^{u}}(x)\right]\subseteq[0,1],\left[s_{\phi_{\acute{A}}^{l}}(x),s_{\phi_{\acute{A}}^{u}}(x)\right]\subseteq[0,1],\left[s_{\psi_{\acute{A}}^{l}}(x),s_{\psi_{\acute{A}}^{u}}(x)\right]\subseteq[0,1]$

and $0\le\left(s_{\rho_{\acute{A}}^{u}(x)}\right)^{2}+\left(s_{\phi_{\acute{A}}^{u}(x)}\right)^{2}+\left(s_{\psi_{\acute{A}}^{u}(x)}\right)^{2}\le 1,\ \forall x\in\chi$.

For particular x, this IVLSFS becomes IVLSFN.

Property 4.4.2. Some basic properties of IVLSFN:

Let $s_{a'^{1}}=\left\langle\left[s_{a_1},s_{b_1}\right],\left[s_{c_1},s_{d_1}\right],\left[s_{e_1},s_{f_1}\right]\right.$, $s_{a'^{2}}=\left\langle\left[s_{a_2},s_{b_2}\right],\left[s_{c_2},s_{d_2}\right],\left[s_{e_2},s_{f_2}\right]\right\rangle$,

and $s_{a'^{3}}\left(\left[s_{a_3},s_{b_3}\right],\left[s_{c_3},s_{d_3}\right],\left[s_{e_3},s_{f_3}\right]\right)$ be three IVLSFN and $\lambda,\lambda_1,\lambda_2\ge 0$

be three scalars. Then, we have the following properties:

(i) $s_{a'^{1}}\cup s_{a'^{2}}=\left\langle\left[\max\left\{s_{a_1}s_{a2}\right\},\max\left\{s_{b_1}s_{b_2}\right\}\right],\left[\min\left\{s_{c_1}s_{c_2}\right\},\min\left\{s_{d_1}s_{d_2}\right\}\right]\right.$,

$\left.\left[\min\left\{s_{e_1}s_{e_2}\right\},\min\left\{s_{f_1}s_{f_2}\right\}\right]\right\rangle$

(ii) $s_{a'^{1}}\cap s_{a'^{2}}=\left\langle\left[\min\left\{s_{a_2}s_{a_2}\right\},\min\left\{s_{b_1}s_{b_2}\right\}\right],\left[\max\left\{s_{c_1}s_{c_2}\right\},\max\left\{s_{d_1}s_{d_2}\right\}\right]\right.$,

$\left.\left[\max\left\{s_{e_1}s_{e_2}\right\},\max\left\{s_{f_1}s_{f_2}\right\}\right]\right\rangle$

(iii) $$s_{a'1} \oplus s_{a'2} = \left\langle \left[\left(\left(s_{a_1}\right)^2 + \left(s_{a_2}\right)^2 - \left(s_{a_1}\right)^2 + \left(s_{a_2}\right)^2 \right)^{\frac{1}{2}}, \left(\left(s_{b_1}\right)^2 + \left(s_{b_2}\right)^2 - \left(s_{b_1}\right)^2 \left(s_{b_2}\right)^2 \right)^{\frac{1}{2}}; \right. \right.$$

$$\left[s_{c_1} s_{c_2},\ s_{d_1} s_{d_2} \right], \left[\left(\left(1 - \left(s_{a_2}\right)^2 \left(s_{e_1}\right)^2\right) + \left(1 - \left(s_{a_1}\right)^2 \left(s_{e_2}\right)^2 - \left(s_{e_1}\right)^2 \left(s_{e_2}\right)^2 \right)^{\frac{1}{2}} \right) \right.$$

$$\left. \left. \left(\left(1 - \left(s_{b_2}\right)^2\right) \left(s_{f_1}\right)^2 + \left(1 - \left(s_{b_1}\right)^2\right) \left(s_{f_2}\right)^2 - \left(s_{f_1}\right)^2 \left(s_{f_2}\right)^2 \right)^{\frac{1}{2}} \right] \right\rangle$$

(iv) $$s_{a'1} \otimes s_{a'2} = \left\langle \left[s_{a_1} s_{a_2},\ s_{b_1} s_{b_2} \right] \left[\left(\left(s_{c_1}\right)^2 + \left(s_{c_2}\right)^2 - \left(s_{c_1}\right)^2 + \left(s_{c_2}\right)^2 \right)^{\frac{1}{2}},\ \left(\left(s_{d_1}\right)^2 + \left(s_{d_2}\right)^2 - \right. \right. \right.$$

$$\left. \left(s_{d_1}\right)^2 \left(s_{d_2}\right)^2 \right)^{\frac{1}{2}}, \left[\left(\left(1 - \left(s_{c_2}\right)^2 \left(s_{e_1}\right)^2\right) + \left(1 - \left(s_{c_1}\right)^2 \left(s_{e_2}\right)^2 - \left(s_{e_1}\right)^2 \left(s_{e_2}\right)^2 \right)^{\frac{1}{2}} \right), \right.$$

$$\left. \left. \left(\left(1 - \left(s_{d_2}\right)^2\right) \left(s_{f_1}\right)^2 + \left(1 - \left(s_{d_1}\right)^2\right) \left(s_{f_2}\right)^2 - \left(s_{f_1}\right)^2 \left(s_{f_2}\right)^2 \right)^{\frac{1}{2}} \right] \right\rangle$$

(v) $$s_{a'1}^{\lambda} = \left\langle \left[s_{a_1}^{\lambda}, s_{b_1}^{\lambda} \right], \left[\left(1 - \left(1 - s_{c_1}^2\right)^{\lambda}\right)^{\frac{1}{2}}, \left(1 - \left(1 - s_{d_1}^2\right)^{\lambda}\right)^{\frac{1}{2}} \right], \right. ;$$

$$\left. \left[\left(\left(1 - s_{c_1}^2\right)^{\lambda} - \left(1 - s_{c_1}^2 - s_{e_1}^2\right)^{\lambda} \right)^{\frac{1}{2}}, \left(\left(1 - s_{d_1}^2\right)^{\lambda} - \left(1 - s_{d_1}^2 - s_{f_1}^2\right)^{\lambda} \right)^{\frac{1}{2}} \right] \right\rangle$$

(vi) $$\lambda s_{a'1} = \left\langle \left[\left(1 - \left(1 - s_{a_1}^2\right)^{\lambda}\right)^{\frac{1}{2}}, \left(1 - \left(1 - s_{b_1}^2\right)^{\lambda}\right)^{\frac{1}{2}} \right], \left[s_{c_1}^{\lambda}, s_{d_1}^{\lambda} \right], \right. ;$$

$$\left. \left[\left(\left(1 - s_{a_1}^2\right)^{\lambda} - \left(1 - s_{a_1}^2 - s_{e_1}^2\right)^{\lambda} \right)^{\frac{1}{2}}, \left(\left(1 - s_{b_1}^2\right)^{\lambda} - \left(1 - s_{b_1}^2 - s_{f_1}^2\right)^{\lambda} \right)^{\frac{1}{2}} \right] \right\rangle$$

(vii) $s_{a'1} \oplus s_{a'2} = s_{a'2} \oplus s_{a'1}$;

(viii) $s_{a'1} \otimes s_{a'2} = s_{a'2} \otimes s_{a'1}$;

(ix) $\lambda\left(s_{a'^1} \oplus s_{a'^2}\right) = \lambda s_{a'^1} \oplus \lambda s_{a'^2}$;

(x) $\lambda_1 s_{a'^1} \oplus \lambda_2 s_{a'^1} = \left(\lambda_1 + \lambda_2\right) s_{a'^1}$;

(xi) $\left(s_{a'^1} \otimes s_{a'^2}\right)^{\lambda} = \left(s_{a'^1}\right)^{\lambda} \otimes \left(s_{a'^2}\right)^{\lambda}$;

(xii) $\left(s_{a'^1}\right)^{\lambda_1} \otimes \left(s_{a'^1}\right)^{\lambda_2} = \left(s_{a'^1}\right)^{\lambda_1+\lambda_2}$.

All of the above calculations on IVLSFNs are done based on the suffixes of IVLSFN following Property 4.3.11.

Example 4.3. For example, $s_{a'^1} = \left\langle\left[s_{0.80}, s_{0.85}\right], \left[s_{0.20}, s_{0.25}\right], \left[s_{0.10}, s_{0.15}\right]\right\rangle$ is an IVLSFN. Since, in linguistic form, suffixes are used for all mathematical calculations, then $s_{a'^1}$ can be expressed as: $s_{a'^1} = \left\langle[0.80, 0.85], [0.20, 0.25], [0.10, 0.15]\right\rangle$ Similarly, the other elements are expressed.

4.5 PROPOSED RANKING APPROACH ON IVLSFS

In this section, we define a new ranking approach on IVLSFN.

Definition 4.5.1. Consider $s_{a'^1} = \left\langle\left[s_{a_1}, s_{b_1}\right], \left[s_{c_1}, s_{d_1}\right], \left[s_{e_1}, s_{f_1}\right]\right\rangle$ an IVLSFN. Then its de-spherifuzzificated value is defined as:

$$\mathfrak{D}(s_{a'^1}) = \theta\left|\text{diam. }\left(0.5\left[s_{a_1}, s_{b_1}\right] + 0.3\left[s_{c_1}, s_{d_1}\right] - 0.2\left[s_{e_1}, s_{f_1}\right]\right)\right|; \quad (4.3)$$

$$= \theta\left|\text{diam. }\left(0.5[a_1, b_1] + 0.3[c_1, d_1] - 0.2[e_1, f_1]\right)\right|; \quad (4.4)$$

$$= \theta\left|\text{diam. }\left(\left[0.5a_1 + 0.3c_1 - 0.2f_1, 0.5b_1 + 0.3d_1 - 0.2e_1\right]\right)\right|. \quad (4.5)$$

where diam. indicates the diameter of the corresponding interval and θ is an arbitrary constant used for scale change.

Example 4.5.2. For a IVLSFN $s_{a'} = \left\langle\left[s_{0.85}, s_{0.95}\right], \left[s_{0.10}, s_{0.15}\right], \left[s_{0.05}, s_{0.10}\right]\right\rangle$, we have $\mathfrak{D}\left(s_{a'}\right) = 0.075$ for $\theta = 1$.

Definition 4.5.3. Let $s_{a'1} = \left\langle \left[s_{a_1}, s_{b_1}\right], \left[s_{c_1}, s_{d_1}\right], \left[s_{e_1}, s_{f_1}\right] \right\rangle$, and $s_{a'2} = \left\langle \left[s_{a_2}, s_{b_2}\right], \left[s_{c_2}, s_{d_2}\right], \left[s_{e_2}, s_{f_2}\right] \right\rangle$ be two IVLSFNs. Then,

(i) $s_{a'1} \succ s_{a'1}$ if $\mathfrak{D}\left(s_{a'1}\right) > \mathfrak{D}\left(s_{a'1}\right)$;

(ii) $s_{a'1} \prec s_{a'1}$ if $\mathfrak{D}\left(s_{a'1}\right) < \mathfrak{D}\left(s_{a'1}\right)$;

(iii) $s_{a'1} = s_{a'1}$ if $\mathfrak{D}\left(s_{a'1}\right) = \mathfrak{D}\left(s_{a'1}\right)$.

Theorem 4.5.4. $\mathfrak{D}\left(s_{a'}\right)$ is linear.

Proof: This proof is straightforward and left for the reader.

4.6 ALGORITHM OF WARSHALL–FLOYD-BASED SHORTEST ROUTE FINDING PROBLEM

The Warshall–Floyd algorithm is used to find the shortest path between all nodes. The algorithm works on an n-node system depicted in a $n \times n$ square matrix. Every element of the matrix gives the distance *dij* from node *i* to *j*. When *i* is connected directly to j, this *dij* is finite and infinite elsewhere. The logic of the Warshall–Floyd algorithm is simple. Assume three nodes, *i*, *j*, and *k*, with the weights (distances) on the three edges. It is shorter to reach *k* from *i*, passing through *j* if $d_{ij} + d_{jk} < d_{ik}$. In this situation, it is optimal to replace the direct route from *i* to *k* rather than the route via *j*. This triangle operation is applied to the shortest path finding problem. The Floyd algorithm follows the following proposition:

Proposition 6.1. Considering a triangle operation for successive values $k = 1,2,\ldots,n$, each entry *dij* becomes equal to the length of the shortest path from *i* to *j*, assuming the distances of each path are positive.

Proof: This proposition is easy to prove using mathematical induction principles.

Next, we define the stepwise algorithmic structure to solve the shortest path problem.

Algorithm 4.1: Solving the shortest path problem using the Warshall–Floyd algorithm.

Step 1: Consider the network problem to find out the shortest route;

Step 2: Fabricate the decision matrix D^0 corresponding to the set of distances and the node sequence matrix S^0. The diagonal elements are given with '–' to indicate that they are blocked since there is no edge from a node to itself. Set $k = 1$;

Step 3: Assume row k and column k as pivot row and pivot column. Apply the triangle operation to each entry d_{ij} in D^{k-1}, $\forall i, j$.

If $d_{jk} + d_{kj} < d_{ij}(i \neq k, j \neq k, i \neq j)$ then,

Construct D^k by replacing d_{ij} in D^{k-1} with $d_{ik} + d_{kj}$;

Construct S^k by replacing s_{ij} in S^{k-1} with k;

Step 4: Set $k = k + 1$ and repeat Step 3 until there is no change in triangle operations;

Step 5: Optimal results (paths) are obtained after n steps using D^n and S^n, as follows:

d_{ij} is the shortest distance within i and j from D^n;

From S^n, determine the intermediate node $k = S_{ij}$ that gives the path $i \to k \to j$. Stop, if $S_{ij} = k$, $S_{kj} = j$; all the intermediate nodes of the path are evaluated. Otherwise, repeat the process in between i and k and j and k.

4.7 CASE STUDIES: COVID-19 AND TRANSPORTATION SYSTEMS IN WEST BENGAL OF INDIA

In West Bengal, India, local trains are not running regularly due to COVID-19 precautionary action taken by the Government of West Bengal by its order. A few special trains are running, but these are not adequate, according to passengers. So, passengers are using bus services generally. In our problem, we consider the routes between two towns, namely, Midnapore and Garhbeta. Both of these towns are connected by train and bus services. Passengers from Midnapore to Garhbeta use bus routes that pass through several stoppages and stands. In the course of travel, they are asked for fares according to the distances. These fares are determined by the conductors of local buses after bargaining or negotiating with passengers. So, fares have three parts: available mostly, not available, and clearly not determined. These characters are the same as membership degrees, nonmembership degrees, and hesitance degrees of spherical fuzzy sets. Since the conversation in the time of fair allotment is performed through linguistic mode in between passengers and conductors, we consider the linguistic spherical fuzzy environment. Incorporating interval structures gives the data a more realistic form. In the following table (Table 4.1), we show the seven-term IVLSFNs with descriptions.

TABLE 4.1 Interval-Valued Linguistic Spherical Fuzzy Numbers with Descriptions.

Descriptions of distances using fares	Interval-valued linguistic spherical fuzzy numbers
Absolutely more high	$\langle [s_{0.45}, s_{0.48}], [s_{0.10}, s_{0.15}], [s_{0.50}, s_{0.53}] \rangle$
Very high	$\langle [s_{0.60}, s_{0.65}], [s_{0.20}, s_{0.22}], [s_{0.20}, s_{0.23}] \rangle$
High	$\langle [s_{0.66}, s_{0.71}], [s_{0.25}, s_{0.30}], [s_{0.25}, s_{0.30}] \rangle$
Moderate	$\langle [s_{0.72}, s_{0.75}], [s_{0.20}, s_{0.30}], [s_{0.05}, s_{0.10}] \rangle$
Slightly moderate	$\langle [s_{0.76}, s_{0.80}], [s_{0.20}, s_{0.30}], [s_{0.45}, s_{0.50}] \rangle$
Low	$\langle [s_{0.80}, s_{0.85}], [s_{0.20}, s_{0.30}], [s_{0.10}, s_{0.15}] \rangle$
Very low	$\langle [s_{0.85}, s_{0.95}], [s_{0.10}, s_{0.15}], [s_{0.05}, s_{0.10}] \rangle$

TABLE 4.2 Edges and Distances Represented by IVLSFS (from Figure 4.2).

Edges	Distances represented by IVLSFS	De-spherifuzzificated values
1–2, 2–3, 5–7, 6–8	$\langle [s_{0.60}, s_{0.65}], [s_{0.20}, s_{0.22}], [s_{0.20}, s_{0.23}] \rangle$	7.4
3–4, 10–11	$\langle [s_{0.45}, s_{0.48}], [s_{0.10}, s_{0.15}], [s_{0.50}, s_{0.53}] \rangle$	7.2
4–5, 8–10	$\langle [s_{0.66}, s_{0.71}], [s_{0.25}, s_{0.30}], [s_{0.25}, s_{0.30}] \rangle$	10
12–13	$\langle [s_{0.72}, s_{0.75}], [s_{0.20}, s_{0.30}], [s_{0.05}, s_{0.10}] \rangle$	11
5–6, 7–9, 9–12	$\langle [s_{0.76}, s_{0.80}], [s_{0.20}, s_{0.30}], [s_{0.45}, s_{0.50}] \rangle$	12
11–12	$\langle [s_{0.80}, s_{0.85}], [s_{0.20}, s_{0.30}], [s_{0.10}, s_{0.15}] \rangle$	13
1–5	$\langle [s_{0.85}, s_{0.95}], [s_{0.10}, s_{0.15}], [s_{0.05}, s_{0.10}] \rangle$	15

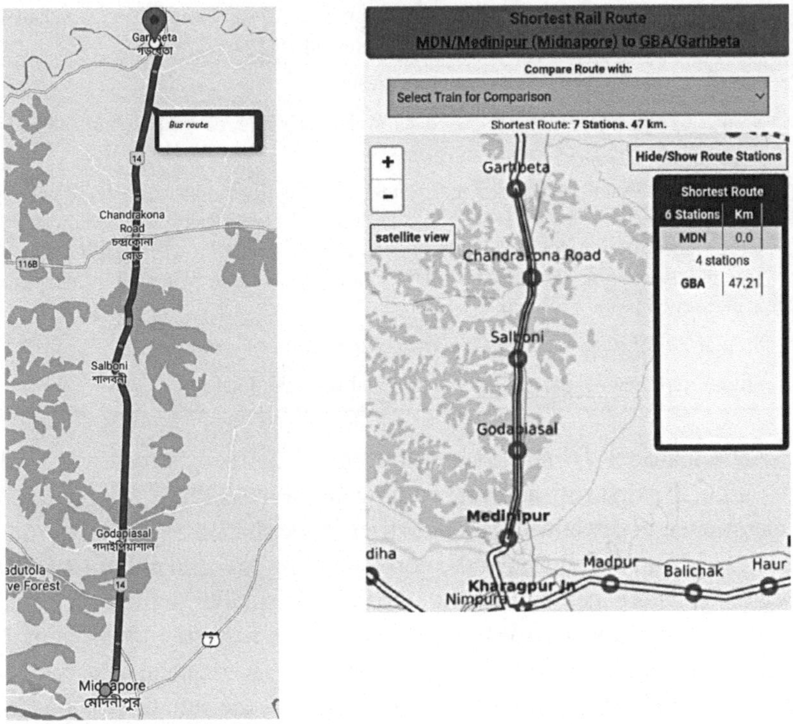

(a) **Bus route** (b) **Train route**

FIGURE 4.1 Bus and train routes between Midnapore and Garhbeta.

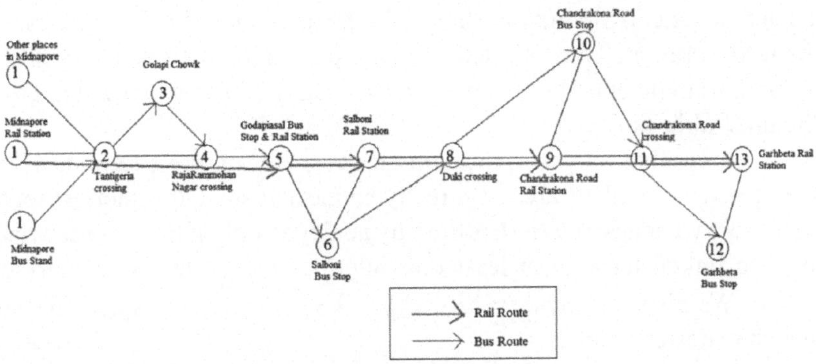

FIGURE 4.2 A picture of the possible routes between Midnapore and Garhbeta.

Algorithm 4.2: Solving the shortest path problem using the modified Warshall–Floyd algorithm through the proposed ranking approach of IVLSFN.
Step 1: Consider the network problem given;
Step 2: Construct the distance-based decision matrix after de-spherifuzzification of each edge value using eqs 4.3–4.5;
Step 3: Construct the node sequence matrix S^0 as given subsequently. The diagonal elements are marked with '–' to indicate that they are blocked. In addition, we define a big transportation cost 'M' corresponding to those edges where path exists but no medium is used for transportation. This modification of the Warshall–Floyd algorithm is performed in our problem;
Step 4: Perform the Warshall–Floyd algorithm using Algorithm 4.1;
Step 5: Optimal results (paths) are obtained from the final D^n and S^n.

The matrices D^0 and S^0 give the initial network structures. In the first iteration, the first row and first column are pivoted for D^0. Apply the triangle operation to every entry (except entries in the shaded region). For this, select any entry, find the entries in the corresponding row and column of the shaded region, and take their sum. If the sum value is smaller than the selected entry, at that point, box it so it could be supplanted by this lower sum, leaving others unaltered. For example, in D^0, entry d_{25} is changed as it has 7.4 in the row and 15 in the column of the shaded region, the sum of which is 22.4. So, 'M' of d_{25} is replaced by 22.4 in the next iteration. Similar changes occur in entry d_{52}. Because this is Iteration 1, the corresponding entries of node matrix S^0 are replaced by 1.

Consider the next iteration after incorporating the changes from the first iteration; see D^1 and S^1. The second row and column, being pivoted, are shaded now. Again, apply the triangle operation to all entries in the unshaded region and make corresponding changes in the node matrix to have matrices for the next iteration. Thus, all iterations are performed until no changes are made using triangle operations. Thus, performing the intermediate steps, we get the final table as D^{12}.

Here, we get that the transportation fare cost from node 1 to node 13 is 46.4. But nodes 1 and 13 are not directly connected since the shortest route between any two nodes, say (i, j), is directly achieved only if $S_{ij} = j$; otherwise, i and j are linked through at least one intermediate node. Here, we see $s_{1\to13} = 9$. We also see, $s_{9\to13} = 13$, $s_{7\to9} = 9, s_{5\to7} = 7$, and $s_{1\to5} = 5$. So, we get the shortest route:

$$1 \to 5 \to 7 \to 9 \to 13,$$

and this indicates the train routes from Figures 4.1 and 4.2. But due to the government's circular of stopping the local trains, people use the following path from Midnapore to Garhbeta and vice-versa:

$1 \rightarrow 2 \rightarrow 3 \rightarrow 4 \rightarrow 5 \rightarrow 6 \rightarrow 8 \rightarrow 10 \rightarrow 11 \rightarrow 12 \rightarrow 13$. This results in high transportation costs, and these long path distances spread more viruses from virus carriers.

TABLE 4.3 D^0.

	1	2	3	4	5	6	7	8	9	10	11	12	13
1	–	7.4	M	M	15	M	M	M	M	M	M	M	M
2	7.4	–	7.4	M	M	M	M	M	M	M	M	M	M
3	M	7.4	–	7.2	M	M	M	M	M	M	M	M	M
4	M	M	7.2	–	10	M	M	M	M	M	M	M	M
5	15	M	M	10	–	12	7.4	M	M	M	M	M	M
6	M	M	M	M	12	–	M	7.4	M	M	M	M	M
7	M	M	M	M	7.4	M	–	M	12	M	M	M	M
8	M	M	M	M	M	7.4	M	–	M	10	M	M	M
9	M	M	M	M	M	M	12	M	–	M	M	M	12
10	M	M	M	M	M	M	M	10	M	–	7.2	M	M
11	M	M	M	M	M	M	M	M	M	7.2	–	13	M
12	M	M	M	M	M	M	M	M	M	M	13	–	11
13	M	M	M	M	M	M	M	M	12	M	M	11	–

TABLE 4.4 S^0.

	1	2	3	4	5	6	7	8	9	10	11	12	13
1	–	2	3	4	5	6	7	8	9	10	11	12	13
2	1	–	3	4	[redacted]	6	7	8	9	10	11	12	13
3	1	2	–	4	5	6	7	8	9	10	11	12	13
4	1	2	3	–	5	6	7	8	9	10	11	12	13
5	1	[redacted]	3	4	–	6	7	8	9	10	11	12	13
6	1	2	3	4	5	–	7	8	9	10	11	12	13
7	1	2	3	4	5	6	–	8	9	10	11	12	13
8	1	2	3	4	5	6	7	–	9	10	11	12	13
9	1	2	3	4	5	6	7	8	–	10	11	12	13
10	1	2	3	4	5	6	7	8	9	–	11	12	13
11	1	2	3	4	5	6	7	8	9	10	–	12	13
12	1	2	3	4	5	6	7	8	9	10	11	–	13
13	1	2	3	4	5	6	7	8	9	10	11	12	–

TABLE 4.5 D^1.

	1	2	3	4	5	6	7	8	9	10	11	12	13
1	–	7.4	M	M	15	M	M	M	M	M	M	M	M
2	7.4	–	7.4	M	22.4	M	M	M	M	M	M	M	M
3	M	7.4	–	7.2	M	M	M	M	M	M	M	M	M
4	M	M	7.2	–	10	M	M	M	M	M	M	M	M
5	15	22.4	M	10	–	12	7.4	M	M	M	M	M	M
6	M	M	M	M	12	–	M	7.4	M	M	M	M	M
7	M	M	M	M	7.4	M	–	M	12	M	M	M	M
8	M	M	M	M	M	7.4	M	–	M	10	M	M	M
9	M	M	M	M	M	M	12	M	–	M	M	M	12
10	M	M	M	M	M	M	M	10	M	–	7.2	M	M
11	M	M	M	M	M	M	M	M	M	7.2	–	13	M
12	M	M	M	M	M	M	M	M	M	M	13	–	11
13	M	M	M	M	M	M	M	M	12	M	M	11	–

TABLE 4.6 S^1.

	1	2	3	4	5	6	7	8	9	10	11	12	13
1	–	2	3	4	5	6	7	8	9	10	11	12	13
2	1	–	3	4	1	6	7	8	9	10	11	12	13
3	1	2	–	4	5	6	7	8	9	10	11	12	13
4	1	2	3	–	5	6	7	8	9	10	11	12	13
5	1	1	3	4	–	6	7	8	9	10	11	12	13
6	1	2	3	4	5	–	7	8	9	10	11	12	13
7	1	2	3	4	5	6	–	8	9	10	11	12	13
8	1	2	3	4	5	6	7	–	9	10	11	12	13
9	1	2	3	4	5	6	7	8	–	10	11	12	13
10	1	2	3	4	5	6	7	8	9	–	11	12	13
11	1	2	3	4	5	6	7	8	9	10	–	12	13
12	1	2	3	4	5	6	7	8	9	10	11	–	13
13	1	2	3	4	5	6	7	8	9	10	11	12	–

TABLE 4.7 D^2.

	1	2	3	4	5	6	7	8	9	10	11	12	13
1	–	7.4	14.8	M	15	M	M	M	M	M	M	M	M
2	7.4	–	7.4	M	22.4	M	M	M	M	M	M	M	M
3	14.8	7.4	–	7.2	29.8	M	M	M	M	M	M	M	M
4	M	M	7.2	–	10	M	M	M	M	M	M	M	M
5	15	22.4	29.8	10	–	12	7.4	M	M	M	M	M	M
6	M	M	M	M	12	–	M	7.4	M	M	M	M	M
7	M	M	M	M	7.4	M	–	M	12	M	M	M	M
8	M	M	M	M	M	7.4	M	–	M	10	M	M	M
9	M	M	M	M	M	M	12	M	–	M	M	M	12
10	M	M	M	M	M	M	M	10	M	–	7.2	M	M
11	M	M	M	M	M	M	M	M	M	7.2	–	13	M
12	M	M	M	M	M	M	M	M	M	M	13	–	11
13	M	M	M	M	M	M	M	M	12	M	M	11	–

TABLE 4.8 S^2.

	2	3	4	5	6	7	8	9	10	11	12	13	
1	–	2	2	[redacted]	5	6	7	8	9	10	11	12	13
2	1	–	3	[redacted]	1	6	7	8	9	10	11	12	13
3	2	2	–	4	2	6	7	8	9	10	11	12	13
4	[redacted]		3	–	5	6	7	8	9	10	11	12	13
5	1	1	2	4	–	6	7	8	9	10	11	12	13
6	1	2	3	4	5	–	7	8	9	10	11	12	13
7	1	2	3	4	5	6	–	8	9	10	11	12	13
8	1	2	3	4	5	6	7	–	9	10	11	12	13
9	1	2	3	4	5	6	7	8	–	10	11	12	13
10	1	2	3	4	5	6	7	8	9	–	11	12	13
11	1	2	3	4	5	6	7	8	9	10	–	12	13
12	1	2	3	4	5	6	7	8	9	10	11	–	13
13	1	2	3	4	5	6	7	8	9	10	11	12	–

TABLE 4.9 D^{12}.

	1	2	3	4	5	6	7	8	9	10	11	12	13
1	–	7.4	14.8	22	15	27	22.4	34.4	34.4	44.4	51.6	57.4	46.4
2	7.4	–	7.4	14.6	22.4	34.4	29.8	41.8	41.8	51.8	59	64.8	53.8
3	14.8	7.4	–	7.2	17.2	29.2	24.6	36.6	36.6	46.6	53.8	59.6	48.6
4	22	14.6	7.2	–	10	22	17.4	29.4	29.4	39.4	46.6	52.4	41.4
5	15	22.4	17.2	10	–	12	7.4	19.4	19.4	29.4	36.6	42.4	31.4
6	27	34.4	29.2	22	12	–	19.4	7.4	31.4	17.4	24.6	37.6	43.4
7	22.4	29.8	24.6	17.4	7.4	19.4	–	26.8	12	36.8	44	35	24
8	34.4	41.8	36.6	29.4	19.4	7.4	26.8	–	38.8	10	17.2	30.2	41.2
9	34.4	41.8	36.6	29.4	19.4	31.4	12	38.8	–	43.2	36	23	12
10	44.4	51.8	46.6	39.4	29.4	17.4	36.8	10	43.2	–	7.2	20.2	31.2
11	51.6	59	53.8	46.6	36.6	24.6	44	17.2	36	7.2	–	13	24
12	57.4	64.8	59.6	52.4	42.4	37.6	35	30.2	23	20.2	13	–	11
13	46.4	53.8	48.6	41.4	31.4	43.4	24	41.2	12	31.2	24	11	–

TABLE 4.10 S^{12}.

	1	2	3	4	5	6	7	8	9	10	11	12	13
1	–	2	2	3	5	5	5	6	7	8	10	12	9
2	1	–	3	3	5	5	5	6	7	8	10	12	9
3	2	2	–	4	4	5	5	6	7	8	10	12	9
4	3	3	3	–	5	5	5	6	7	8	10	12	9
5	1	1	4	4	–	6	7	6	7	8	10	12	9
6	5	5	5	5	5	–	7	8	7	8	10	11	9
7	5	5	5	5	5	6	–	6	9	8	10	12	9
8	6	6	6	6	6	6	6	–	7	10	10	11	12
9	7	7	7	7	7	7	7	7	–	12	12	12	13
10	8	8	8	8	8	8	8	8	8	–	11	11	12
11	10	10	10	10	10	10	10	10	10	10	–	12	12
12	11	11	11	11	11	11	11	11	11	11	11	–	13
13	1	2	3	4	5	6	7	12	9	12	12	12	–

4.7.1 COMPARATIVE ANALYSIS

In this section, we compare our study with others in the literature.

If we consider the score and accuracy function approach (Kahraman and Gu¨ndo˘gdu, 2021), that is, for $s_{a'^1} = \langle [s_{a_1}, s_{b_1}], [s_{c_1}, s_{d_1}], [s_{e_1}, s_{f_1}] \rangle$, the score and accuracy functions are defined respectively,

$$Score(s_{a'^1}) = \frac{(s_{a_1})^2 + (s_{b_1})^2 - (s_{c_1})^2 - (s_{d_1})^2 - (s_{e_1}/2)^2 - (s_{f_1}/2)^2}{2}$$

$$Accu(s_{a'^1}) = \frac{(s_{a_1})^2 + (s_{b_1})^2 + (s_{c_1})^2 + (s_{d_1})^2 + (s_{e_1}/2)^2 + (s_{f_1})^2}{2}$$

we get the same shortest route in between nodes 1 and 13 as: $1 \to 5 \to 7 \to 9 \to 13$.

If we consider the score function approach (Kahraman and Gündoğdu, 2021), that is, for $s_{a'^1} = \langle [s_{a_1}, s_{b_1}], [s_{c_1}, s_{d_1}], [s_{e_1}, s_{f_1}] \rangle$, the score function is defined as:

$$Score(s_{a'^1}) = \frac{(s_{a_1})^2 + (s_{b_1})^2 - (s_{c_1})^2 - (s_{d_1})^2 - (s_{e_1})^2 - (s_{f_1})^2}{3}$$

we get the same shortest route in between nodes 1 and 13, that is, $1 \to 5 \to 7 \to 9 \to 13$.

4.8 CONCLUSION

In this chapter, we have introduced a modified version of the Warshall-Floyd algorithm to find out the shortest route between two towns in India, and we have discussed here the government's wrong decision stopping the local trains where train routes are the shortest routes. In this chapter, we have encountered a spherical fuzzy set under interval-valued linguistic information. Our study highlights that the public transit system is a critical infrastructural system for a community during pandemic situation like COVID-19.

The main advantage of our proposed method is its simple calculation and the assumption of the linguistic format of the considered variables, whereas the main disadvantage of this method is its time consumption during calculation. Finally, there is an urgent need for attitudinal and behavioral surveys and further analysis of administrative decisions of transportation systems during a pandemic, and this study can help to design effective public transit systems using the shortest routes among different places to minimize the spread of viruses.

CONFLICT OF INTEREST

The author would like to announce that there is no conflict of interest.

FUNDING

No funding.

KEYWORDS

- **shortest path problem**
- **Warshall–Floyd algorithm**
- **linguistic set**
- **spherical fuzzy number**
- **interval-valued linguistic spherical fuzzy number**
- **COVID-19 transportation system**

REFERENCES

1. Ashraf, S.; Abdullah, S. Spherical Aggregation Operators and Their Application in Multi-Attribute Group Decision-Making. *Int. J. Intell. Syst.* **2019,** *34* (3), 493–523.
2. Ashraf, S.; Abdullah, S.; Abdullah, L. Child Development Influence Environmental Factors Determined Using Spherical Fuzzy Distance Measures. *Mathematics* **2019,** *7* (8), 661. https://doi.org/10.3390/math7080661.
3. Ashraf, S.; Abdullah, S.; Mahmood, T. Spherical Fuzzy Dombi Aggregation Operators and Their Application in Group Decision Making Problems. *J. Ambient Intell. Human. Comput.* **2019,** *11*, 2731–**2749,** https://doi.org/10.1007/s12652-019-01333-y.
4. Atanassov, K. T. Intuitionistic Fuzzy Sets. *Fuzzy Sets Syst.* **1986,** *20*, 87–96.

5. Bhaumik, A.; Roy, S. K.; Intuitionistic Interval-Valued Hesitant Fuzzy Matrix Games with a New Aggregation Operator for Solving Management Problem. *Granular Comput* **2019**. https://doi.org/10.1007/s41066-019-00191-5.
6. Bhaumik, A.; Roy, S. K.; Li, D. F. Analysis of Triangular Intuitionistic Fuzzy Matrix Games Using Robust Ranking. *J. Intell. Fuzzy Syst.* **2017,** *33*, 327–336.
7. Bhaumik, A.; Roy, S. K.; Li, D. F. (α, β, γ)-Cut Set Based Ranking Approach to Solving Bi-Matrix Games in Neutrosophic Environment. *Soft Comput.* **2021,** *25*, 2729–2739.
8. Bhaumik, A.; Roy, S. K.; Weber, G. W. Hesitant Interval-Valued Intuitionistic Fuzzy-Linguistic Term Set Approach in Prisoners Dilemma Game Theory Using TOPSIS: A Case Study on Human-trafficking. *Central Eur. J. Oper. Res.* **2020,** *28*, 797–816.
9. Bhaumik, A.; Roy, S. K.; Weber, G. W. Multi-Objective Linguistic-Neutrosophic Matrix Game and Its Applications to Tourism Management. *J. Dyn. Games* **2020**. https://doi.org/10.3934/jdg.2020031
10. Bliss, L. When the World Stops Moving, 2020. https://www.citylab.com/transportation/2020/03/coronavirus-impact-public-transit-street-traffic-data-trains/607915/
11. Boltu¨rk, E. AS/RS Technology Selection Using Spherical Fuzzy TOPSIS and Neutrosophic TOPSIS. In: *International Conference on Intelligent and Fuzzy Systems*; Springer: Cham, 2019; pp 969–976.
12. Chen, N.; Xu, Z. S. Properties of Interval-Valued Hesitant Fuzzy Sets. *J. Intell. Fuzzy Syst.* **2014,** *27*, 143–158.
13. Chen, N.; Xu, Z. S.; Xia, M. M. Interval-Valued Hesitant Preference Relations and Their Applications to Group Decision Making. *Knowl.-Based Syst.* **2013,** *37*, 528–540.
14. Gu¨ndoˇgdu, F. K. Principals of Spherical Fuzzy Sets. In: *International Conference on Intelligent and Fuzzy Systems*; Springer: Cham, 2019; pp 15–23.
15. Gu¨ndoˇgdu, F. K.; Kahraman, C. A Novel Spherical Fuzzy Analytic Hierarchy Process and Its Renewable Energy Application. *Soft Comput.* **2019,** 1–15. https://doi.org/10.1007/s00500-019-04222-w.
16. Gu¨ndoˇgdu, F. K.; Kahraman, C. A Novel VIKOR Method Using Spherical Fuzzy Sets and Its Application to Warehouse Site Selection. *J. Intell. Fuzzy Syst.* **2019,** *37* (1), 1197–1211.
17. Gu¨ndoˇgdu, F. K.; Kahraman, C. Spherical Fuzzy Analytic Hierarchy Process (AHP) and Its Application to Industrial Robot Selection. In: *International Conference on Intelligent and Fuzzy Systems*; Springer: Cham, 2019; pp 988–996.
18. Gu¨ndoˇgdu, F. K.; Kahraman, C. Spherical Fuzzy Sets and Spherical Fuzzy TOPSIS Method. *J. Intell. Fuzzy Syst.* **2019,** *36* (1), 337–352.
19. Herrera, F.; Herrera-Viedma, E. Linguistic Decision Analysis: Steps for Solving Decision Problems Under Linguistic Information. *Fuzzy Sets Syst.* **2000,** *115*, 67–82.
20. Herrera, F.; Herrera-Viedma, E.; Martinez, L. A Fuzzy Linguistic Methodology to Deal with Unbalanced Linguistic Term Sets. *IEEE Trans. Fuzzy Syst.* **2008,** *16*, 354–370.
21. Herrera, F.; Herrera-Viedma, E.; Verdegay, J. L. A Sequential Selection Process in Group Decision Making with Linguistic Assessment. *Inf. Sci.* **1995,** *85*, 223–239.
22. Jiang, X. P.; Wei, G. W. Some Bonferroni Mean Operators with 2-Tuple Linguistic Information and Their Application to Multiple Attribute Decision Making. *J. Intell. Fuzzy Syst.* **2014,** *27*, 2153–2162.

23. Jin, H.; Ashraf, S.; Abdullah, S.; Qiyas, M.; Bano, M.; Zeng, S. Linguistic Spherical Fuzzy Aggregation Operators and Their Applications in Multi-Attribute Decision Making Problems. *Mathematics* **2019,** *7* (5), 413.
24. Kacprzyk, J.; Zadrozny, S. Linguistic Database Summaries and Their Protoforms; Towards Natural Language Based Knowledge Discovery Tools. *Inf. Sci.* **2005,** 173 (4), 281–304.
25. Kahraman, C.; Gu¨ndo˘gdu, F. K., Eds. Decision Making with Spherical Fuzzy Sets. *Stud. Fuzziness Soft Comput.* **2021,** 392.
26. Krishnakumari, P.; Cats, O. Virus Spreading in Public Transport Networks. Working paper. Transport Institute TU Delft, April 30, 2020.
27. Liu, L.; Miller, H. J.; Scheff, J. The impacts of COVID-19 pandemic on public transit demand in the United States. PLoS ONE, 15 (11):e0242476. https://doi.org/10.1371/journal.pone.0242476
28. Liu, P.; Khan, Q.; Mahmood, T.; Hassan, N. T-Spherical Fuzzy Power Muirhead Mean Operator Based on Novel Operational Laws and Their Application in Multi-Attribute Group Decision Making. *IEEE Access* **2019,** *7*, 22613–22632.
29. Liu, P.; Zhu, B.; Wang, P. A Multi-Attribute Decision-Making Approach Based on Spherical Fuzzy Sets for Yunnan Baiyaos R&D Project Selection Problem. *Int. J. Fuzzy Syst.* **2019,** *21* (7), 2168–2191.
30. Liu, P.; Zhu, B.; Wang, P.; Shen, M. An Approach Based on Linguistic Spherical Fuzzy Sets for Public Evaluation of Shared Bicycles in China. *Eng. App. Artif. Intell.* **2020,** *87*, 103295.
31. Mahmood, T.; Ullah, K.; Khan, Q.; Jan, N. An Approach Toward Decision-Making and Medical Diagnosis Problems Using the Concept of Spherical Fuzzy Sets. *Neural Comput. App.* **2019,** *31* (11), 7041–7053.
32. Martinez, L.; Liu, J.; Yang, J. B.; Herrera, F. A Multi-Granular Hierarchical Linguistic Model for Design Evaluation Based on Safety and Cost Analysis. *Int. J. Intell. Syst.* **2005,** *20* (12), 1161–1194.
33. O¨ zkan, B.; O¨ zceylan, E.; Kabak, M.; Da˘gdeviren, M. Evaluating the Websites of Academic Departments Through SEO Criteria: A Hesitant Fuzzy Linguistic MCDM Approach. *Artif. Intell. Rev.*, 2019. https://doi.org/10.1007/s10462-019-09681-z.
34. Rafiq, M.; Ashraf, S.; Abdullah, S.; Mahmood, T.; Muhammad, S. The Cosine Similarity Measures of Spherical Fuzzy Sets and Their Applications in Decision Making. *J. Intell. Fuzzy Syst.* **2019,** *36* (6), 6059–6073.
35. Rodriguez, R. M.; Martinez, L. An Analysis of Symbolic Linguistic Computing Models in Decision Making. *Int. J. Gen. Syst.* **2013,** *42*, 121–136.
36. Rodriguez, R. M.; Martinez, L.; Herrera, F. Hesitant Fuzzy Linguistic Term Sets for Decision Making. *IEEE Trans. Fuzzy Syst.* **2012,** *20* (1), 109–119.
37. Roy, S. K.; Bhaumik, A. Intelligent Water Management: A Triangular Type-2 Intuitionistic Fuzzy Matrix Games Approach. *Water Resour. Manag.* **2018,** *32*, 949–968.
38. Tirachini, A.; Cats, O. COVID-19 and Public Transportation: Current Assessment, Prospects, and Research Needs. *J. Public Transp.* **2020,** *22* (1), 1–21.
39. Torra, V. Hesitant Fuzzy Sets. *Int. J. Intell. Syst.* **2010,** *25* (6), 529–539.
40. West Bengal Govt. Order. 2020. https://wb.gov.in/upload/MCLNEWS-210628154510621.pdf
41. Xia, M. M.; Xu, Z. S. Hesitant Fuzzy Information Aggregation in Decision Making. *Int. J. Approx. Reason.* **2011,** *52*, 395–407.

42. Xu, Z. S. A Method Based on Linguistic Aggregation Operators for Group Decision Making with Linguistic Preference Relations. *Inf. Sci.* **2004,** *166* (1), 19–30.
43. Xu, Z. S. Deviation Measures of Linguistic Preference Relations in Group Decision Making. *Omega* **2005,** *33*, 249–254.
44. Yager, R. R. Multiple Objective Decision-Making Using Fuzzy Sets. *Int. J. Man Machine. Stud.* **1997,** *9*, 375–382.
45. Yu, D. J.; Zhang, W. Y.; Xu, Y. J. Group Decision Making Under Hesitant Fuzzy Environment with Application to Personnel Evaluation. *Knowl.-Based Syst.* **2013,** *52*, 1–10.
46. Zadeh, L. A. Fuzzy Sets. *Inf. Control* **1965,** *8* (3), 338–353.
47. Zadeh, L. A. The Concept of a Linguistic Variable and Its Application to Approximate Reasoning-1. *Inf. Sci.* **1975,** *8*, 199–249.

CHAPTER 5

An Intelligent Agent-Based Approach for COVID-19 Patient Distribution Management

CLAUDIU-IONUT POPIRLAN[1], ADRIANA BURLEA SCHIOPOIU[2], CRISTINA POPIRLAN[1], and AHMED A. ELNGAR[3,4]

[1]*Department of Computer Science, Faculty of Sciences, University of Craiova, Romania*

[2]*Department of Management, Marketing, Business Administration, Faculty of Economy and Business Administration, University of Craiova, Romania*

[3] *Faculty of Computers and Artificial Intelligence, Beni-Suef University, Beni-Suef City, Egypt*

[4]*College of Computer Information Technology, American University in the Emirates, United Arab Emirates*

ABSTRACT

The provision of COVID-19 medical care typically involves a number of individuals, located in a number of different institutions and situations, whose decisions and actions need to be coordinated if the care is to be effective and efficient. To facilitate this decision-making and to ensure that the coordination process runs smoothly, the use of Artificial Intelligence through intelligent agents is becoming increasingly widespread. To this end, this research work describes an intelligent agent's system that was implemented to help manage the care process in real-world settings. The intelligent agents themselves are developed using an open-source platform for peer-to-peer

Global Digital Transformation and the Covid-19 Pandemic.
Rajdeep Chowdhury, K. Martin Sagayam, Neyara Radwan, Ahmad A. Elngar (Eds.)

software agent-based applications, called JADE-JAVA Agent Development Framework, which combines a number of intelligent agent techniques: a symbolic decision procedure for decision-making with incomplete and conflicting information, a concept of negotiation and accountability for task allocation, the notions of commitments and conventions for managing coherent cooperation, and a set of communication primitives for inter-agents interactions. The utility of this approach is demonstrated through the development of an application prototype for the clinical management process of COVID-19 treatment.

5.1 INTRODUCTION

Intelligent agents' systems,[12–14] in particular, and Artificial Intelligence,[9] in general, are assuming an increasingly important role in medicine for assisting clinical staff in making decisions under uncertainty (e.g., pandemic crisis, therapy, test selection, and diagnosis decisions). Furthermore, many medical procedures now involve several individuals, in a number of specialist institutions (or departments), whose decisions and actions need to be coordinated if the care is to be effective and efficient (state-of-the-art).

In the last few years, many types of research were realized in the sphere of AI-related to Covid-19 based on segmentation from computed tomography (CT) images.[1,5]

The pandemic crisis provided to worldwide that is very important to have a quick reaction to an invisible enemy and more than ever it is necessary to provide the healthcare sector with flexible applications for the clinical management process of COVID-19 treatment. Li et al.[2] developed an application used to diagnose COVID-19 and differentiate it from the other forms of pneumonia. Wang et al.[3] elaborated a generally applicable methodology composed of a library of models, training tools process for data collection, annotation, testing, user interaction design, and clinical deployment. Their multi-modal model can be continuously improved by adding new data that will increase the detection accuracy even though CT image is one of the factors for the rapid diagnosis.

Shaban et al.[4] made an analysis of the three types of COVID-19 diagnosis (e.g., RT-PCR test, CT test, and numerical laboratory tests) and elaborated their Hybrid Diagnose Strategy (HDS) based on preprocessing, feature ranking, and classification phase.

Ferguson et al.[6] considered two strategies as important for fighting against this enemy: suppression (by reducing the reproduction of the virus) and mitigation (by vaccine or medication). Aggarwal et al.[8] based on a critical literature review of the COVID-19 arrived at the conclusion that multi-criterion decision-making progress is suitable to make the best decisions under pressure.

Recently, Gupta et al.[7] compared the behavior of the pandemic crisis with the previous healthcare crisis (e.g., Ebola, MERS, SARS, and h1n1) and they elaborated on a deep learning model that depends on the data received from the source.

This limitation of the model is also a limitation for many AI applications because without high reliability the results are not relevant for diagnosis.

We will try to diminish this limitation and build a prototype able to measure in real time and reduce the uncertainty of epidemiological appearances of COVID-19.

5.2 INTELLIGENT AGENT'S ARCHITECTURE PROPOSAL FOR COVID-19 PATIENT DISTRIBUTION MANAGEMENT

The intelligent agent architecture comprises multiple layers of knowledge, a working memory, a communications manager, and a human–computer interface, as is shown in Figure 5.1. To be efficient in this domain, the intelligent agents need to exhibit both deliberative behavior (e.g., task decomposition, plan selection, and task allocation) and reactive behavior (e.g., respond in a timely manner to the arrival of new data, to changes in existing data, and to varying agent commitments). Within the proposed architecture the deliberative behavior is achieved by the incorporation of decision rules for plan selection, task management rules for task decomposition and allocation, and cooperation rules for formulating commitments. Reactive behavior is achieved by the control layer which responds to changes in the working memory: the arrival of new task results, goals, or messages or changes in existing data, agent commitments, or task states.

The three layers of knowledge which form the key part of the intelligent agent's architecture are as follows:

- Domain knowledge—includes, for example, a knowledge base covering specific medical domains such as the COVID-19 pandemic, a knowledge base of clinical management plans (known as clinical protocols,[8] a database of patient records, and a database of resource availability.

- Inference knowledge—in the form of generic, declarative inference rules which specify inference relations between domain knowledge, existing patient information, and possible new data. Inference rules represent the core of the agent architecture and are subdivided into those for decision-making under uncertainty, those for task management, and those for managing agent cooperation.
- Control knowledge—applies the inference knowledge to the domain knowledge in order to generate new inferences whenever new data is added to the working memory. Logically, this layer is a meta-level which controls the execution of inference rules and domain facts.

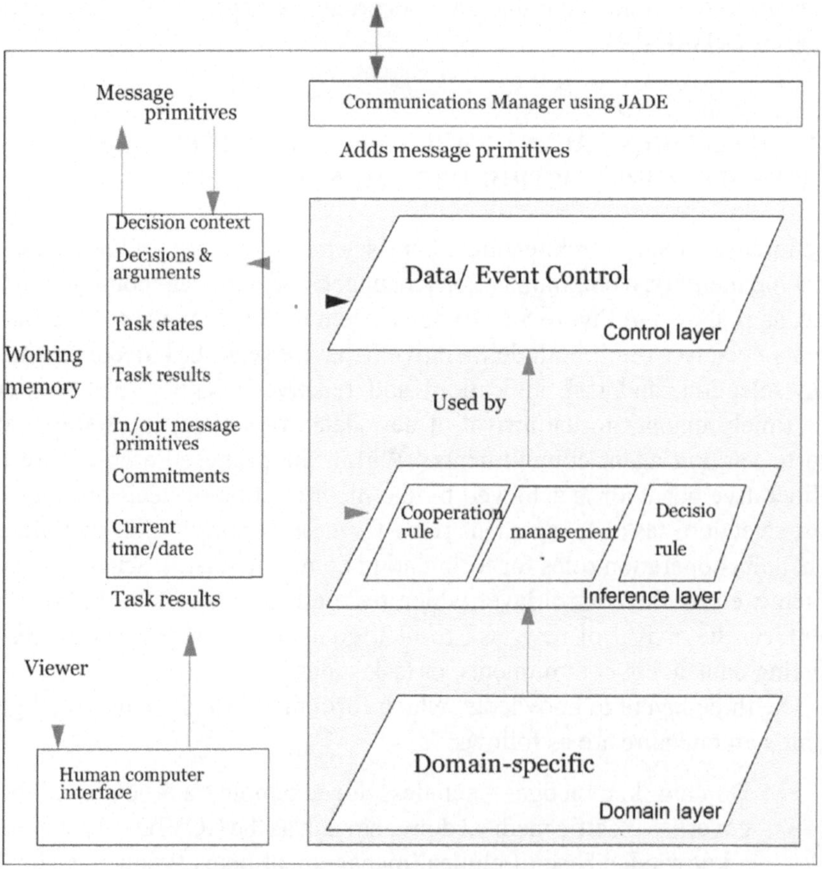

FIGURE 5.1 Intelligent agents' architecture for COVID-19 patient distribution management.

In more detail, the domain knowledge base simply states information and facts about the domain. It says nothing about how the knowledge is to be used. For example, it states that the second stage of the COVID-19 protocol for treating patients contains three subtasks: "admit the patient to hospital," "administer drugs and monitor patient," and "discharge patient":

component ("COVID-19 stage 2," "admit patient to hospital")

component ("COVID-19 stage 2," "administer drugs and monitor patient")

component ("COVID-19 stage 2," "discharge patient")

The inference knowledge base contains rules (implemented as declarative schemas) that specify the inference relations between domain-level knowledge and possibly new information. For example, the following inference schema specifies that the state of a task

becomes "started" once the state of one of its subtasks becomes "started":

schema(conditions(component(Task, Subtask) and state(Subtask, started)), conclusions(state(Task, started)))

In the context of COVID-19 protocol, the above schema implies that the task "COVID-19 stage 2" should become "started" when one of its subtasks (e.g., "admit the patient to the hospital") becomes 'started."

However, it is only at the control level that the actual execution of the inference rules is carried out and new data is added to the working memory:

If schema (Conditions, Conclusions) and

all true (Conditions)

then

add (Conclusions)

For example, within the context of OVID-19 protocol, once the data state ("admit the patient to hospital," started) is asserted to the working memory, the above control rule applies the given inference schema and domain knowledge to add a new piece of data into the working memory: state ("CT 1 stage 2," started).

Bringing all of this together, a sample working session of an intelligent agent is as follows. The agent first specifies an initial goal of finding an appropriate protocol for treating a particular patient's COVID-19 infection. The goal triggers the control layer to apply the decision rules, medical domain knowledge, and patient case data to arrive at a decision (i.e., a suggested treatment protocol called "COVID-19 protocol") and associated arguments. The oncologist endorses that decision and requests the machine to assist him in managing the execution of the protocol. This will, in turn, trigger the control layer to apply the task management rules to decompose the protocol into constituent tasks, propose task allocations to appropriate agents, and generate the corresponding communication primitives. Once the intelligent agent accepts the proposal, the communications manager will convert the primitives into complete messages and send them to the specified agents. Once the chosen agents inform the intelligent agent of their acceptance of the task requests, they and the intelligent agent enter into a dynamic, cooperating agent community. Agent commitments are generated and monitored by triggering the control layer to apply cooperation rules. Cooperation among the agents continues until the entire protocol is in a terminable state (e.g., completed or abandoned).

In the meantime, the oncologist may enter into another agent community and accept task requests from other agents.

There are two main reasons for adopting this functional and logical separation of the domain, inference, and control knowledge. First, it simplifies the representation, reuse, and maintenance of knowledge. Inference knowledge for decision-making, task management, and cooperation can be represented independently of medical domains and can therefore be reused; control knowledge is represented independently of the inference knowledge and so the same control rules can be applied to the three different groups of inference rules. Furthermore, modifications to domain knowledge can be made independently of inference and control knowledge. The second main reason for such a separation is that it provides a convenient basis for knowledge elicitation: domain knowledge can be acquired and modified independently of inference and control knowledge.

An intelligent agent's architecture working memory stores temporary data generated by the control layer, the user, or the communications manager. Examples of the types of information which need to be stored include goals to be achieved, control states of tasks that are currently active, results of completed tasks, incoming and outgoing messages, and current commitments.

Its function is like that of a blackboard, on which new information (or any change which triggers reactions by the control layer) can be added.

The communications manager composes the messages to be sent to the other agents from the primitives produced by firing task management or cooperation rules. It also converts messages which arrive from other agents into primitives that may be used by the cooperation manager.

The human–computer interface defines a scheme for interaction between the support system and its user. The approach is as follows: the computer can perform various functions (i.e., decision-making, task management, communication, and cooperation), but it may not act autonomously on all these capabilities. In general, the computer informs the user of the results of its inferences, and the user must then endorse or authorize them before they can be communicated to external agents. For example, the system may recommend to the intelligent agent that he asks a particular nurse to perform the drug administration subtask; however, the oncologist may have a personal preference for another nurse and may, therefore, be unwilling to make such a referral. In this case, the system will not send an electronic request to the original nurse, but it will instead offer the intelligent agent an alternative solution.

5.2.1 A CLINICAL SCENARIO OF COVID-19 PATIENT DISTRIBUTED MANAGEMENT

When a doctor in a COVID-19 hospital must treat a patient infected, the first decision he must make concerns about the treatment plan which will be adopted. To assist him in making this decision, the intelligent agent consults one of his decision support systems. This system has a built-in decision procedure that is able to deal with incomplete or intuitively inconsistent information, such as evidence in favor of a choice and evidence against the choice. Having weighed the pros and cons of using various treatment options, the system recommends the use of a particular COVID-19 protocol called the "COVID-19 protocol." The intelligent agent authorizes the use of this protocol and requests the computer system to support him in carrying out the treatment.

The COVID-19 protocol consists of a number of treatment stages. According to the protocol, stage 2 is decomposed into a sequence of three subtasks: "admit the patient to hospital," "administer drugs and monitor patient," and "discharge patient." The support system recommends that the oncologist carries out the first task because it knows that he is formally responsible for the admission of his patients. This recommendation is endorsed by the oncologist and consequently, he takes on the role of managing and actually performing the activity. On the recommendation of his support system, the oncologist then decomposes "admit the patient to hospital" into two parallel subtasks: "allocate bed" and "obtain patient consent." The machine recommends, and the intelligent agents accept, that "allocate bed" should be performed by the hospital's resource management department and "obtain patient consent" should be carried out by the intelligent agent. With respect to the former subtask, the intelligent agent sends an electronic request to the resource department to see whether they are willing to take on the responsibility for performing it. Assuming they are and that the task is successfully completed, the patient will be allocated a bed. Once a bed is available and the patient agrees to be admitted to the hospital, the support system recommends that the task "administer drugs and monitor patient" is allocated to a hospital nurse. Assuming she accepts, the protocol dictates that the task should be decomposed into the sequential subtasks of "obtain drug," "administer drug," and "observe patient"—all of which the nurse takes responsibility for. She may subsequently decompose the "observe patient" subtask still further: for example, into "measure body temperature," "take blood samples," and "analyze blood samples" and this decomposition may well involve generating a request to a laboratory to test the patient indicators, such as white blood cell count. However, for the sake of simplicity, this level of decomposition is not described here. Finally, the machine recommends that the third subtask of stage 2, "discharge patient," and its two subtasks, "Instruct Patient" and "Inform" should be carried out by the intelligent agent. In this scenario, information is transferred according to the following pattern: the resource management department must inform the oncologist about the outcome of "allocate bed" (i.e., either that a bed has been allocated as requested or that no bed is available for the requested date) and the nurse has to inform the oncologist of the results of "administer drugs and monitor patient" (e.g., drug has been administered and all patient indicators are normal). Accompanying this information exchange is a concomitant flow of control, in terms of commitments and expectations, between the agents: the oncologist expects the resource management department to perform the activity "allocate bed" once it has been agreed to, similarly he expects the nurse to perform the "administer drugs and monitor patient" task on time once she has consented to execute it.

5.2.2 COMMUNICATION MANAGEMENT

After an extensive analysis of the interactions which can occur in cooperative care organizations, a set of communication primitives, based on speech act theory, have been defined (Table 5.1). Each primitive has a type (illocutionary force) and a content (propositional content), as well as a certain effect on the receiver (perlocutionary force). Having a well-defined set of primitives is important because it means that the ambiguity in message interchange is substantially reduced—each primitive has a clear meaning and must be responded to in a predictable way.

TABLE 5.1 Communication Management.

Type	Content	Effect on receiver
Request	Task; [provisional schedule]; priority: urgent or not; response by the date	Receive Agent evaluates whether to accept the request and informs Send Agent of the decision. If the recipient decides to accept the request, it becomes committed to the task.
Accept	Task; [accepted schedule]	Receive Agent knows Send Agent is committed to the request and that Send Agent will inform it of the outcome of executing the task. Send Agent becomes the contractor for the task and Receive Agent the manager.
Reject	Task; [provisional schedule]	Receive Agent has to request someone else to perform the task on the provisional schedule.
Alter	Task; provisional schedule; acceptable schedule.	Receive Agent to evaluate the acceptable schedule and decide whether to replace the provisional schedule with the acceptable schedule. If so, it sends Send Agent a new request. Otherwise, Receive Agent has to send the original request to someone else.
Propose	Task; [proposed schedule]	Receive Agent may or may not adopt the proposal.
Inform	Any information: data, domain knowledge, or partial plans.	Receive Agent may use the information for local problem solving.
Query	A question: what, how, whether, and so on.	Receive Agent must answer the query, possibly involving extensive local problem solving (e.g., diagnosis and investigation). A reply may be of the type "propose." It may also be "informed," possibly giving the answer "unknown" to the query.
Cancel	Any message of the above types.	Receive Agent should ignore the earlier message.
Acknowledge	Any message of the above types. All messages need to be acknowledged except acknowledgment messages themselves.	Receive Agent is aware of the successful transmission of the message.

The primitive's request, accept, reject, and alter is used during the allocation of tasks and the formulation of agent commitments. Using the example from the scenario described before, the intelligent agent may allocate the task "administer drugs and monitor patient" to a nurse by requesting her to perform it during a period of 10 days starting on the following day. The nurse may accept the task exactly as specified by the intelligent agent, or she may reject it because she is too busy during the next few days (insufficient resources to honor the commitment). Alternatively, the nurse may indicate that she cannot start the task the following day, but she could start it 2 days later (alter).

A suggested act may be the result of a query—for instance, suggesting treatment protocol COVID-19 after being asked how to treat breast cancer. Inform usually follows an accepted request to perform a certain task and is mainly used to disseminate results. Inform may also accompany a request to provide relevant information for the contractor. The fact that cancel is included as a primitive type is because in certain circumstances agents may modify their commitments. Finally, all messages must be acknowledged.

Knowledge about the semantics of the different types of primitives is incorporated into the task manager and cooperation rules, which can dynamically generate message primitives when executed by the control layer. The communications manager then converts these primitives into complete, structured messages using a communication protocol that defines the syntax of inter-agent messages. PRIMITIVE_CONTENT is as defined in Table 5.1:

<message>::= <sender> <receiver> <date> <time> <patient> <transaction primitive>*

<sender>::= <sender name> <contact address>

<sender name>::= <first name> <surname>

<first name>::= NAME

<surname>::= NAME

<contact address>::= <email address> | <postal address>| <telephone number> | <fax number>

<email address>::= EMAIL_ADDRESS

<postal address>::= POSTAL_ADDRESS

<mobile number>::= NUMBER

<receiver>::= <receiver name> <contact address>

<receiver name>::= <first name> <surname>

<date>::= <day> <month> <year>

<day>::= NUMBER

<month>::= NUMBER

<year>::= NUMBER

<time>::= <hour> <minute>

<minute>::= NUMBER

<hour>::= NUMBER

<patient>::= <patient name> <date of birth>

<patient name>::= <first name> <surname>
<date_of_birth>::= <year> <month> <day>
<transaction primitive>::= <primitive type> <primitive content>
<primitive type>::= REQUEST | ACCEPT | REJECT | ALTER | PROPOSE | INFORM |
QUERY | CANCEL | ACKNOWLEDGE
<primitive content>::= PRIMITIVE_CONTENT.

5.3 INTELLIGENT AGENTS TASK MANAGEMENT

Once the decision procedure has selected a particular clinical protocol to achieve the agent's goal, the task manager component is responsible for its decomposition into subtasks, the allocation of subtasks to appropriate agents, and the management of task state transitions. Each of these activities is described in turn in the remainder of this section.

The structure of a generic clinical plan (e.g., COVID-19 protocol) is determined by experts in authority and is precisely defined in a clinical protocol. The task management rules decompose such a protocol into subtasks according to the predefined plan structure (as described in Section 2). Subtasks at the bottom of the plan hierarchy may be primitive actions for humans or machines to perform (such as "allocate bed" and "administer drugs") or they may be decision tasks (such as choosing the right drug for a patient). In the latter case, the decision procedure is used to perform such a task.

To facilitate task allocation, there are two roles associated with each (sub) task within the system—there is one agent who manages the execution of the task (i.e., ensures that it gets executed by somebody within the system and that the result of the execution is sent back to the originator) and one agent who is actually responsible for performing the task (the contractor). Task allocation is, therefore, the process by which the manager of a task finds the most appropriate contractor to perform it. The key structure in our intelligent agent's architecture for making such decisions is that of accountability. Accountability is a static relationship that defines for what and to whom an agent is responsible. It is expressed by the following relation: accountable (Agent1, Agent2, Task Type) which means that Agent1 is accountable to Agent2 for performing tasks of type Task Type. For example, a hospital nurse may be accountable to one or more doctors for monitoring patient data such as temperature and blood pressure. The task manager component uses its accountability relations, together with the generic inference rule given below, to pick the most appropriate contractor for a given task. The underlined term "request" represents a primitive which is sent to the communications manager when this task management rule is fired.

IF Task is necessary &
Task is of type Task Type &
Acquaintance is accountable to the Agent for tasks of
Task Type &
Agent prefers to interact with Acquaintance
concerning Task Type
THEN request (Agent, Acquaintance, perform (Task))

All tasks within our intelligent agent's architecture have a state (either scheduled, canceled, started, completed, or abandoned). The management of the transitions between these states needs to be carefully controlled by the agents because such transitions need to be documented in patients' care records: for example when a task was scheduled, when it was started, when it was completed, and when (why) it was abandoned. Transition management is complicated by the nested structure of the care plans. For example, the following two task management rules specify that when a composite task is canceled, its started subtasks become abandoned and the subtasks that are scheduled but not yet started become canceled:

schema(conditions(state(Task, cancelled), and
component(Task, Subtask) and
state(Subtask, started)),
conclusions (state(Subtask, abandoned)).
schema(conditions(state(Task, cancelled), and
component (Task, Subtask) and
state(Subtask, scheduled)),
conclusions (state(Subtask, cancelled)).

A distinction is made between the states of canceled and abandoned because corrective action is usually needed for an abandoned task (e.g., when the patient has to stop taking a certain drug that he has already been taking for a period) whereas such an action is not normally necessary for a canceled task.

5.3.1 INTELLIGENT AGENTS COOPERATION MANAGEMENT

In our intelligent agent's architecture, the underlying mechanisms on which cooperative interactions are based are those of commitment (pledge to undertake a specified course of action) and convention (means of monitoring commitments in changing circumstances). The former means that if an agent agrees to undertake a task then it will endeavor to execute it at the appropriate time—this implies both that the agent is able to perform the task and that it has the necessary resources. Conventions are needed because commitments are not irrevocable: agents' circumstances may change between the making and the execution of their commitments, and agreed actions may turn out to be undesirable or even impossible to perform. Conventions, therefore, define the conditions under which an agent can drop its commitments and how to behave with respect to other agents in the cooperating group when such circumstances arise.

Given that cooperation is founded on commitments and conventions, two key issues need to be addressed: (1) what is involved in establishing a commitment? and (2) what type of convention is appropriate for monitoring commitments in the given care organization?

5.3.1.1 ESTABLISHING COMMITMENTS

Accountability alone does not guarantee commitment: to commit to a specified task, an agent must also have the necessary resources (temporal and material) which are required to perform that task. For example, a hospital specialist may be accountable to patients for in-hospital breast cancer treatment, but will not become committed to an actual treatment course on a specific patient until the time (temporal resource) and a bed (material resource) are available to perform the treatment. Although agents know what resources are available to themselves, they do not generally have information about the resources of their acquaintances. Therefore, an agent may have to propose the same task to several acquaintances before an acceptable contractor, and hence commitment can be found (made).

When an agent accepts a request, it becomes committed to performing it (i.e., it takes on the role of the contractor) and informs the manager that the task has been accepted using the following inference rule:

IF Acquaintance is requested by Agent to perform Task &
Acquaintance accountable to Agent for Task Type tasks &
Task is of type Task Type &
Task requires Resources &
Resources are available to Acquaintance
THEN Acquaintance becomes committed to Task, AND

accept (Acquaintance, Task, for (Agent))

Commitment to the role of the contractor also entails an additional responsibility—when the task has been completed the contractor must inform the manager about it and any results which have been generated—again this behavior is encoded in a generic inference rule:

IF Task is completed and it produces Results &
Acquaintance is committed to Agent for Task
THEN inform (Acquaintance, Agent, performed (Task),
results-produced (Task, Results))

Note that in both cases, the underlined term represents primitives sent to the communication manager when the appropriate inference rules are triggered.

5.3.1.2 ADAPTIVE MANAGEMENT OF COMMITMENT CHANGES

In most cases, when an agent commits itself to perform a task then that task will indeed be executed. However, in certain well-defined circumstances, it may be appropriate for an agent to renege upon its commitment. There may be an unforeseen lack of resources (e.g., unrelated emergencies may arise), the need for the task may cease to exist (e.g., because of the unexpected death of the patient), or it may no longer be feasible to execute a given task (e.g., planned chemotherapy may have to be withdrawn because the patient has a high temperature resulting from the toxic effect of the drug). Having detailed the conditions under which commitments can be canceled, the convention must also specify how to manage this change both locally and within the wider context of the cooperating group.

The latter is important because it ensures that the cooperating care agents will behave coherently in the face of dynamic and unpredictable changes in the network.

5.4 JAVA AGENT DEVELOPMENT FRAMEWORK

Multi-agent systems can be realized by using any kind of programming language.[1] In particular, object-oriented languages are considered a suitable means because the concept of the agent is similar to the concept of the object. In fact, agents share many properties with objects such as inheritance, message passing, encapsulation, etc. A list of several agent-oriented languages can be found in Ref. [5]. Software platforms and frameworks are the other key means enabling the development of multiagent systems. Most provide a means to deploy multi-agent systems on different types of hardware and operating systems, usually providing middleware to support their execution and essential operations such as communication and coordination.

Some of these platforms and frameworks have the common goal of providing FIPA-compliant functionalities to support interoperation between different multi-agent systems. Java Agent Development (JADE)[2] is a software platform that provides a basic middleware layer functionality which is independent of the specific application, and which simplifies the realization of distributed applications that exploit the software agent abstraction.[3] A significant merit of JADE is that it implements this abstraction over a well-known object-oriented language, Java, providing a simple and friendly API. JADE was implemented to provide programmers with the following ready-to-use and easy-to-customize core functionalities:

- A fully distributed system inhabited by agents;
- Efficient transport of asynchronous messages via a location-transparent API;
- The platform selects the best available means of communication and, when possible, avoids marshaling/unmarshalling java objects;
- A simple, effective, agent life-cycle management;
- Simple APIs and graphical tools are provided to both locally and remotely manage agent life cycles;
- Support for agent mobility: agent migration is made transparent to communicating agents that can continue or interact even during the migration process;
- A subscription mechanism for agents, and even external applications, that wish to subscribe with a platform to be notified of all platform events, including life-cycle-related events and message exchange events;
- A set of graphical tools to support programmers when debugging and monitoring;
- Full compliance with the FIPA specifications.[14]

The architecture of the multi-agent system that includes the related components is described below and can be visualized in Figures 5.2 and 5.3.

The system must be able to control the intelligent agents as described in Sections 2 and 3.

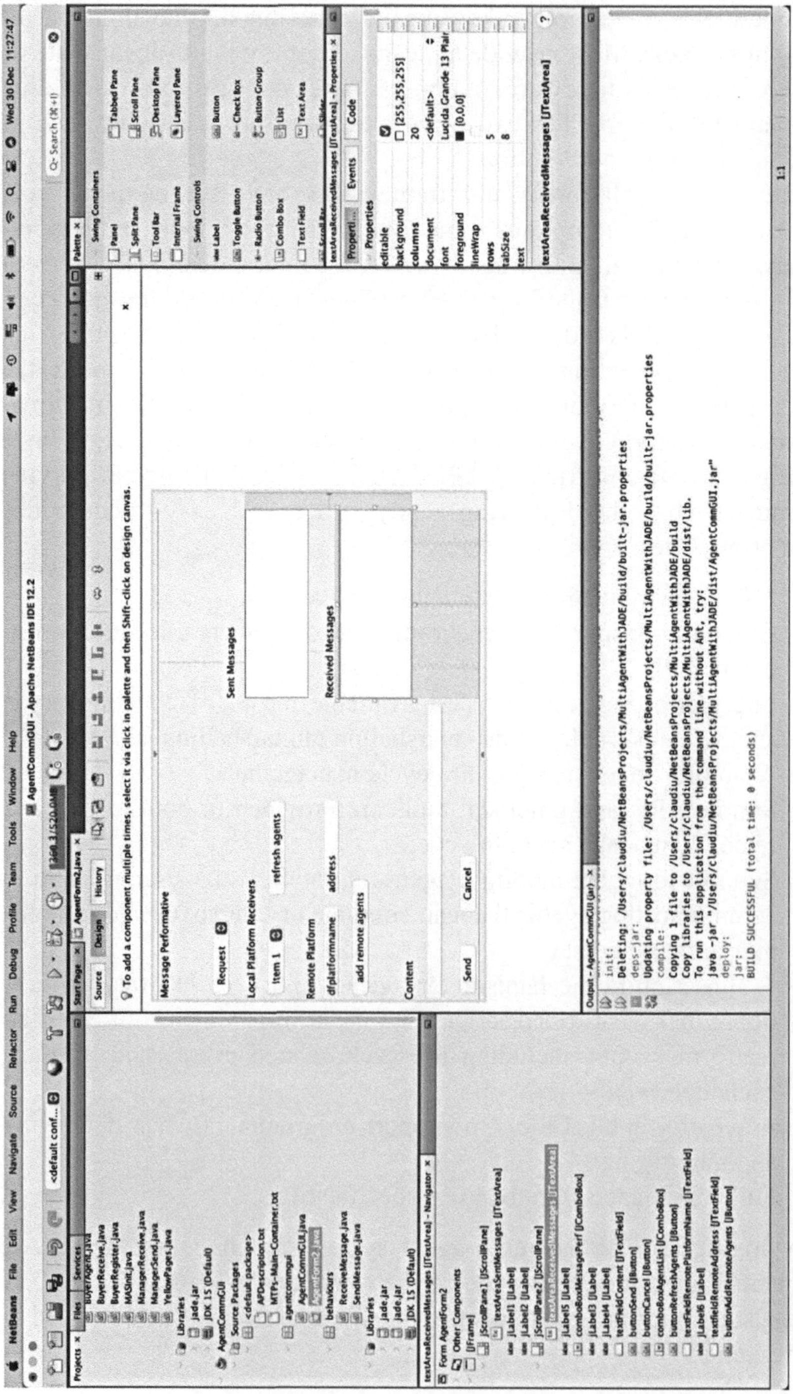

FIGURE 5.2 JADE implementation 1.

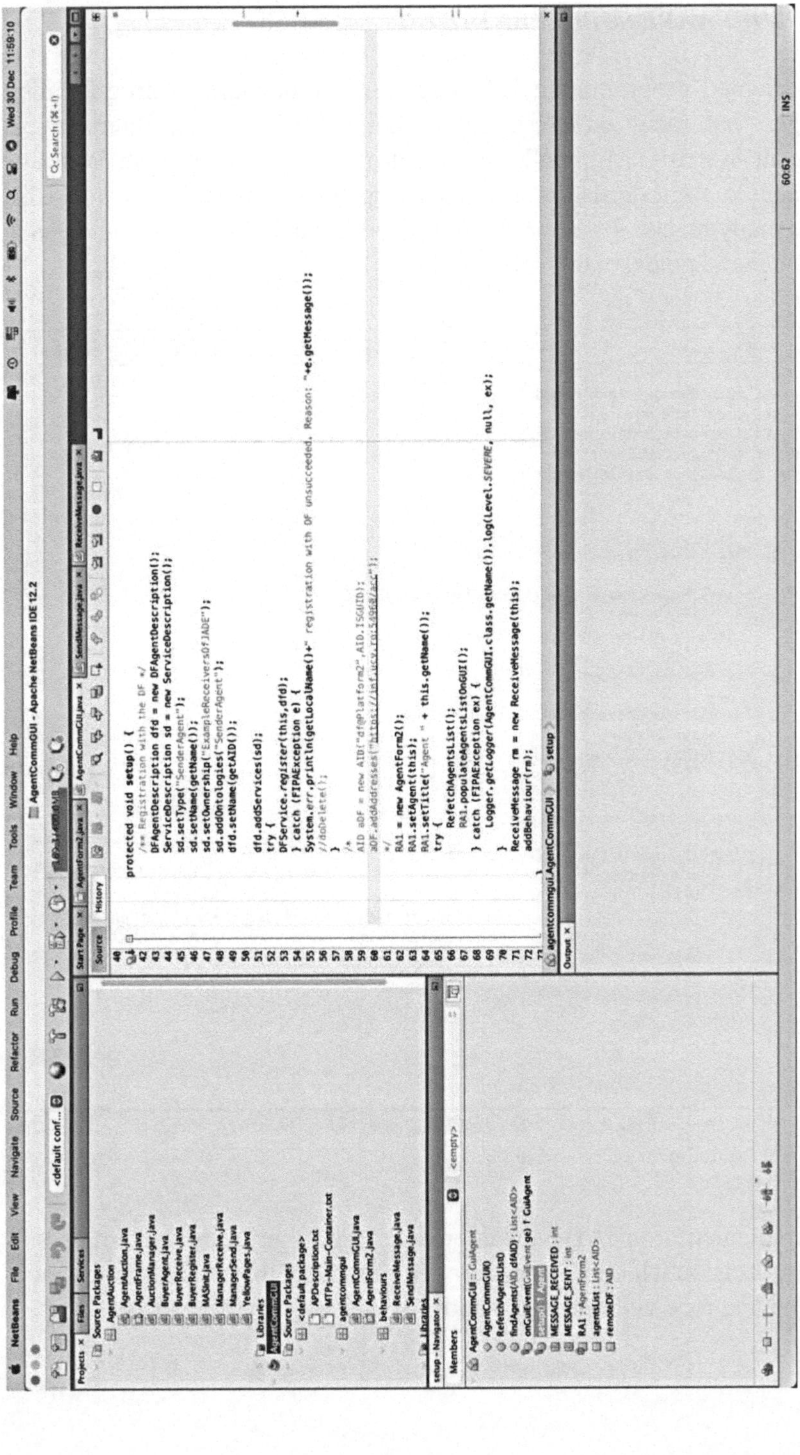

FIGURE 5.3 JADE implementation 2.

5.4.1 SOFTWARE IMPLEMENTATION

Agent-oriented programming languages are a new class of programming languages that focus on taking into account the main characteristics of multiagent systems. Minimally, an agent-oriented programming language must include some structure corresponding to an agent, but it may also provide many mechanisms for supporting additional attributes of an agency such as beliefs, goals, plans, roles, and norms.

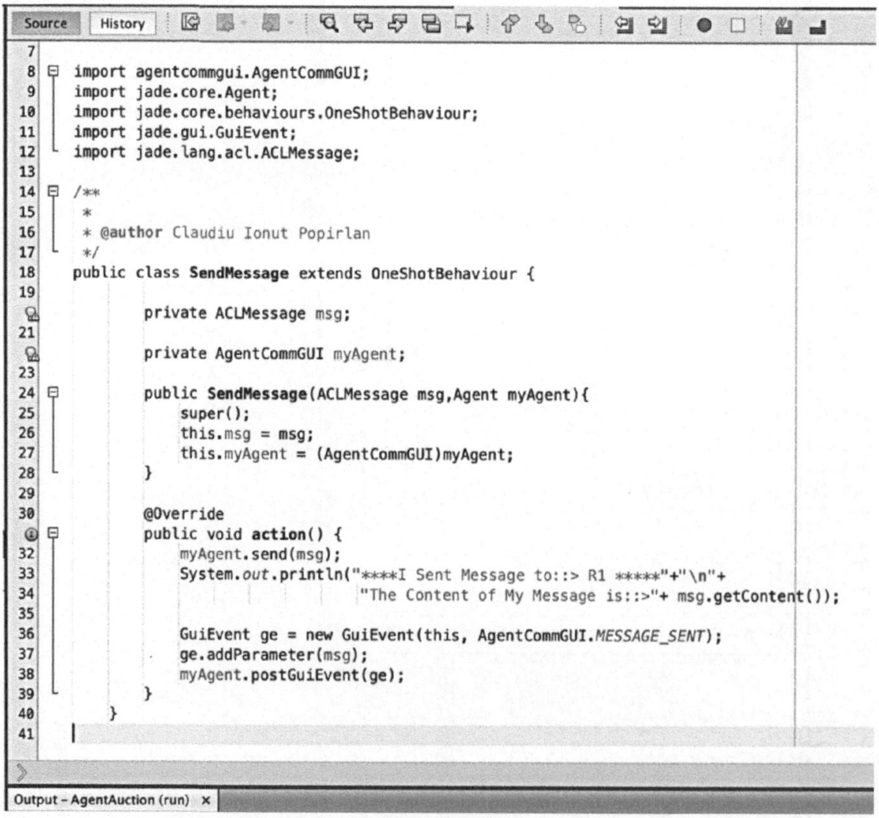

FIGURE 5.4 Intelligent agent—send message.

To implement it in JADE the approach of multi-criterion system uses to take best decision which is necessary defining the communication statement with send and receive Figures 5.4 and 5.5.

```
Source   History
    */
public class ReceiveMessage extends CyclicBehaviour {

        // Variable to Hold the content of the received Message
        private String Message_Performative;
        private String Message_Content;
        private String SenderName;
        private String MyPlan;
        private AgentCommGUI myAgent;

        public ReceiveMessage(Agent myAgent){
            super();
            this.myAgent = (AgentCommGUI) myAgent;
        }

        public void action() {
            ACLMessage msg = myAgent.receive();
            if(msg != null) {

                Message_Performative = msg.getPerformative(msg.getPerformative());
                Message_Content = msg.getContent();
                SenderName = msg.getSender().getLocalName();
                System.out.println(" ****I Received a Message***" +"\n"+
                        "The Sender Name is::>"+ SenderName+"\n"+
                        "The Content of the Message is::> " + Message_Content + "\n"+
                        ":::: And Performative is::> " + Message_Performative + "\n");
                System.out.println("ooooooooooooooooooooooooooooooooooooooo");

                GuiEvent ge = new GuiEvent(this, AgentCommGUI.MESSAGE_RECEIVED);
                ge.addParameter(msg);
                myAgent.postGuiEvent(ge);

            }

        }
    }
Output - AgentAuction (run) ×
```

FIGURE 5.5 Intelligent agent—receive message.

Then, different objective will use many techniques such as questioning based on sets of people, which done by using specialized tools to take a decision. As well as set of decider used to understand the multi-criterion for decision support system based on some parameters and contraries.

```
Output - AgentAuction (run) ×
INFO: Service jade.core.mobility.AgentMobility initialized
Dec 30, 2020 11:29:51 AM jade.core.BaseService init
INFO: Service jade.core.event.Notification initialized
Dec 30, 2020 11:29:51 AM jade.mtp.http.HTTPServer <init>
INFO: HTTP-MTP Using XML parser com.sun.org.apache.xerces.internal.jaxp.SAXParserImpl$JAXPSAXParser
Dec 30, 2020 11:29:51 AM jade.core.messaging.MessagingService boot
INFO: MTP addresses:
http://claudius-mbp:7778/acc
Dec 30, 2020 11:29:51 AM jade.core.AgentContainerImpl joinPlatform
INFO: --------------------------------------
Agent container Container1M@192.168.2.48 is ready.
--------------------------------------------
                                                  AgentAuction (run)        running...
```

FIGURE 5.6 JADE agent container running.

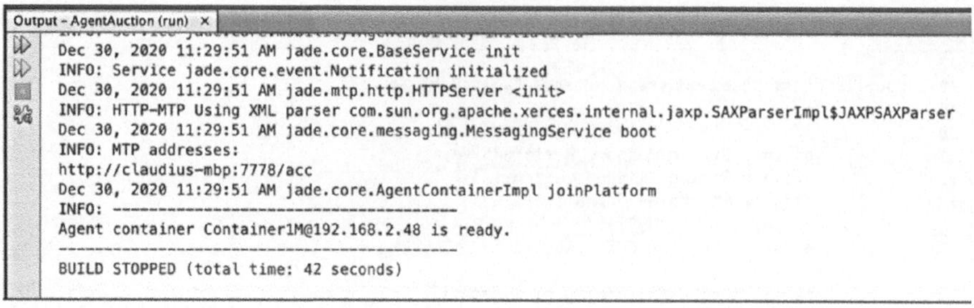

FIGURE 5.7 JADE agent container finish running.

5.4.2 RESULTS COMPARISON

In this section, our Intelligent Agents-based architecture for COVID-19 patient distribution management, developed in JADE, is briefly compared with some of the well-known architectures and systems in the agent literature.

GRATE[7] is also a layered architecture that provides a generic cooperation module, situation assessment module, control module, and application-specific module. However, the GRATE framework lacks an uncertainty management mechanism which is essential for medical decision-making. Also, GRATE's layers are functionally separated rather than logically separated—the additional benefit of this logical separation is that it provides a convenient basis for declarative specification and for logical verification and validation of the various layers of knowledge (e.g., in a formal language such as ML2).[9]

The same two observations can be made of other similar layered architectures such as INTERMAP[10] and Touring Machines.[11]

Coordinator[12] is a conversational system for coordinated action which is based on Searle's speech act theory.[13] However, while the generation and monitoring of speech acts and commitments are centralized in the coordinator, our proposed intelligent agent's architecture distributes both these functions (thus helping to reduce the communication bottleneck). Also, the functionality of the coordinator is limited to coordination alone through the generation of speech acts and commitments, whereas our implemented architecture accommodates additional functions such as a generic decision module for decision-making under uncertainty.

In both cases, the working memory is changed through the application of functionally separated modules of inference rules. However, in addition to functional separation, our proposed intelligent agent's architecture also emphasizes the logical layering of knowledge for the reasons stated above and provides a set of generic knowledge modules for cooperation and decision-making.

5.5 CONCLUSIONS

In this chapter, an Intelligent Agents based approach for COVID-19 patient distribution management in a virtual reality environment (JADE implementation) was presented. The tasks from the COVID-19 protocol stage 2 were controlled using intelligent agents. A very realistic overcome procedure for intelligent agents has been created using JADE, their real behavior (interaction between intelligent agents and the environment) evaluated using Java, and its dynamics and results visualized.

Li et al.[2] designed a hybrid framework by the modification of an original JADE and an original Code with the purpose of increasing both exploration and exploitation ability.

Numerous techniques and systems have been developed in the medical informatics community for tackling isolated aspects of medical decision-making. However, despite a well-documented need for supporting an integrated range of different functions (including uncertainty management, task management, and coordination), there has been very little prior work that attempts to provide comprehensive procedures and integrated decision support for these different aspects of healthcare. For example, Wang et al.[3] proposed a combined *segmentation–classification* model pipeline based on the best diagnosis model by empirically training and evaluating the models within the library, and this model was divided into two stages: 3D segmentation and classification.

A preliminary evaluation of our prototype indicates that in real clinical application settings where exact probabilities and utilities are difficult to obtain the built-in symbolic decision procedure is more effective than conventional numerical methods. This affirmation is in consensus with the results of Badem et al.[23] who proposed a method to solve numerical optimization problems and Umar et al.[25] who statistically analyzed some genetic algorithms, and they elaborated a new computing paradigm. As testing capabilities continue to improve, AI-based tools in general and multi-agent systems, in particular, become more useful to assist in clinical settings rather than to replace medical professionals.

FUNDING STATEMENT

The author(s) received no specific funding for this study.

CONFLICTS OF INTEREST

The authors declare that they have no conflicts of interest to report regarding the present study.

KEYWORDS

- **intelligent agents**
- **COVID-19**
- **JADE simulation**
- **multi-agent systems**
- **patient distribution management**

REFERENCES

1. Chen et al. Deep Learning-Based Model for Detecting 2019 Novel Coronavirus Pneumonia on High-Resolution Computed Tomography. *Sci. Rep.* **2020,** *10*, 19196.
2. Li, L.; Qin, L.; Xu, Z.; Yin, Y.; Wang, X.; Kong, B.; Bai, J.; Lu, Y.; Fang,; Song, Q.; et al. Artificial Intelligence Distinguishes Covid-19 from Community Acquired Pneumonia on Chest CT. *Radiology* **2020,** 200905.
3. Wang B.; et al. AI-Assisted CT Imaging Analysis for COVID-19 Screening: Building and Deploying a Medical AI System. *Appl. Soft Comput.* **2020,** *98*, 106897. https://doi.org/10.1016/j.asoc.2020.106897
4. Shaban, W. M.; Rabie, A. H.; Saleh, A. I.; Abo-Elsoud, M. A. Detecting COVID-19 Patients Based on Fuzzy Inference Engine and Deep Neural Network. *Appl. Soft Comput.* **2020,** *99*, 106906, https://doi.org/10.1016/j.asoc.2020.106906.
5. Wu, D.; et al. Severity and Consolidation Quantification of COVID-19 From CT Images Using Deep Learning Based on Hybrid Weak Labels. *IEEE J. Biomed. Health Inf.* **2020,** *24* (12), 3529–3538. DOI: 10.1109/JBHI.2020.3030224
6. Ferguson, N. M.; Laydon, D.; Nedjati-Gilani, G.; Imai, N.; Ainslie, K.; Baguelin, M.; Bhatia, S.; Boonyasiri, A.; Cucunubá, Z.; CuomoDannenburg, G. Impact of Non-Pharmaceutical Interventions (NPIs) to Reduce COVID-19 Mortality and

Healthcare Demand. [Online] **2020.** https://www.imperial.ac.uk/mrc-global-infectious-disease-analysis/covid19/report-9-impact-of-npis-on-covid-19/ (accessed Dec 28, 2020)

7. Gupta, M.; Jain, R.; Taneja, S.; Chaudhary, G.; Khari, M.; Verdú, E. Real-Time Measurement of the Uncertain Epidemiological Appearances of COVID-19 Infections *Appl. Soft Comput.* **2021,** *101*, 107039. https://doi.org/10.1016/j.asoc.2020.107039
8. Aggarwal, L.; Goswami, P.; Sachdeva, S. Multi-Criterion Intelligent Decision Support System for COVID-19. *Appl. Soft Comput.* **2021,** *101*, 107056. https://doi.org/10.1016/j.asoc.2020.107056.
9. Russell, S.; Norvig, P. In *Artificial Intelligence: A Modern Approach*, Second Edition; Prentice Hall, 2003.
10. Bellifemine, F.; Caire, G.; Poggi, A.; Rimassa, G. JADE: A White Paper. **2003,** *3* (3), 6–19. http://jade.tilab.com/papers/2003/WhitePaperJADEEXP.pdf.
11. Bellifemine, F.; Caire, G.; Trucco, T.; Rimassa, G. JADE Programmer's Guide, JADE 3.3 Edition, Telecom Italia S.p.A., 2005. http://sharon.cselt.it/projects/jade/ .
12. Wooldridge, M. Intelligent Agents. In *Multiagent Systems: a Modern Approach to Distributed Artificial Intelligence*; Weiß, G., Ed.; MIT Press: Cambridge: Massachusetts, 1999; pp 27–77.
13. Wooldridge, M. An Introduction to MultiAgent Systems, 3rd Edition; John Wiley and Sons LTD., 2005.
14. Wooldridge, M.; Jennings, N. R. Intelligent Agents: Theory and Practice. *Knowl. Eng. Rev.* **1995,** *10* (2), 115–152.
15. Jennings, N. R.; Mamdani, E. H.; Laresgoiti, I.; Perez, J.; Corera J. GRATE: A General Framework for Cooperative Problem Solving. *Intell. Syst. Eng.* **1992,** *1* (2), 102–114.
16. Gordon, C.; Herbert, S. I.; Jackson-Smale, A.; Renaud-Salis, J. L. In *Care Protocols and Healthcare Informatics*, Proceedings of Artificial Intelligence in Medicine Europe 93; Munich, Germany, 1993; pp 289–309.
17. van Harmelen, F. ML2: A Formal Language for KADS Models of Expertise. *Knowl. Acquisition* **1992,** *4* (1).
18. Muller, J. P.; Pischel, M.; Thiel, M. A Pragmatic Approach to Modelling Autonomous Interacting Systems: A Preliminary Report. Intelligent Agents. In *Lecture Notes in Artificial Intelligence*; Wooldridge, M. J., Jennings, N. R., Eds.; Springer: Verlag, 1995; vol 890.
19. Ferguson, I. A. Integrated Control and Coordinated Behaviour: A Case for Agent Models in Intelligent Agents. In *Lecture Notes in Artificial Intelligence*; Wooldridge, M. J., Jennings, N. R., Eds.; Springer: Verlag, 1995; vol 890.
20. Winograd, T.; Flores, F. Understanding Computers and Cognition: A New Foundation for Design; Ablex Publishing: Norwood, 1986.
21. Searle, J. R. In *Speech Acts: An Essay in the Philosophy of Language*; Cambridge University Press, 1969.
22. FIPA. Foundation for Intelligent Physical Agents, 2013 [Online]. http://www.fipa.org/.
23. Badem, H.; Basturk, A.; Caliskan, A.; Yuksel, M. E. A New Hybrid Optimization Method Combining Artificial Bee Colony and Limited-Memory BFGS Algorithms for Efficient Numerical Optimization. *Appl. Soft Comput.* **2018,** *70*, 826–844. https://doi.org/10.1016/j.asoc.2018.06.010.
24. Li, G.; Lin, Q.; Cui, L.; Du, Z.; Liang, Z.; Chen, J.; Lu, N.; Ming, Z. A Novel Hybrid Differential Evolution Algorithm with Modified CoDE and JADE. *Appl. Soft Comput.* **2016,** *47*, 577–599. https://doi.org/10.1016/j.asoc.2016.06.011

25. Umar, M.; Sabir, Z.; Raja, M. A. Z. Intelligent Computing for Numerical Treatment of Nonlinear Prey–Predator Models. *Appl. Soft Comput.* **2019,** *80*, 506–524. https://doi.org/10.1016/j.asoc.2019.04.022

CHAPTER 6

An Artificial Intelligence (AI) Approach in Medical Science for Predicting COVID-19 Cases

RANJAN K. MOHAPATRA[1], AJIT BEHERA[2], OM PRAKASH JENA[3], PRADEEP K. DAS[4], SUMAN CHATTERJEE[5], SHARUN KHAN[6], MADHUMITA PAL[7], MUHAMMAD BILAL[8], and KULDEEP DHAMA[9]

[1]*Department of Chemistry, Government College of Engineering, Keonjhar, Odisha, India*

[2]*Department of Metallurgical & Materials Engineering, National Institute of Technology, Rourkela, Odisha, India*

[3]*Department of Computer Science, Ravenshaw University, Odisha, India*

[4]*Department of Chemistry, N. C. (Autonomous) College, Jaipur, Odisha, India*

[5]*Department of Mechanical Engineering, National Institute of Technology, Rourkela, Odisha, India*

[6]*Division of Surgery, ICAR-Indian Veterinary Research Institute, Izatnagar, Bareilly, Uttar Pradesh, India*

[7]*Department of Electrical Engineering, Government College of Engineering, Keonjhar, Odisha, India*

[8]*School of Life Science and Food Engineering, Huaiyin Institute of Technology, Huaian, China*

[9]*Division of Pathology, ICAR-Indian Veterinary Research Institute, Izatnagar, Bareilly, Uttar Pradesh, India*

Global Digital Transformation and the Covid-19 Pandemic.
Rajdeep Chowdhury, K. Martin Sagayam, Neyara Radwan, Ahmad A. Elngar (Eds.)

HIGHLIGHTS

- We have proposed an AI approach using ANN for predicting the incidence of COVID-19 in Odisha.
- The number of days, temperature (day/night), and region's humidity were considered input parameters.
- The cumulative number of deceased cases, confirmed cases, and active cases are contemplated as output responses.
- The proposed model for the network is adequate and will be helpful in providing high-quality healthcare services.

ABSTRACT

Background and Aims: The prediction of COVID-19 cases is attracting continuing interest due to global importance, particularly in the most affected countries. In India, 10,720,048 confirmed cases and 154,010 deaths have been reported as of 29th January 2021. The uncertainty regarding the growth rate of COVID-19 cases makes it difficult for the healthcare system to adapt according to the increasing requirements. Artificial intelligence (AI) and AI-based models can help to rapidly predict and forecast the outcome and impact of outbreaks. Some studies have also been reported for the modeling of the recent pandemic COVID-19 worldwide in different countries using artificial intelligence and machine-learning algorithms.

Methods: Odisha is a major state in India and ranked 11th position as per population. This state is having a population of 4.7 Crores in 2020 as predicted. Now this state has reported 3,34,958 confirmed COVID-19 cases. In the present study, we have proposed an AI approach using an artificial neural network (ANN) for predicting the incidence of COVID-19 in Odisha in the upcoming days. The number of days, the temperature at day and night time, and the region's humidity were considered input parameters. While the cumulative number of deceased cases, confirmed cases, and active cases for COVID-19 in Odisha state are contemplated as output responses for the study.

Results: For the adequacy of the developed model, mean relative error was calculated between testing and predicted data obtained from the model. The mean relative errors of 0.0986, 0.0756, and 0.096 for the number of deceased cases, confirmed cases, and active cases, respectively, show the robustness in predicting the COVID-19 responses by adapting the feed-forward back propagation methodology.

Conclusions: The proposed models for the network are adequate and relevant in predicting COVID-19 cases. The accurate prediction of the disease will help in providing high-quality healthcare services. Therefore, we hope the study may provide useful information for monitoring the COVID-19 outbreak.

6.1 INTRODUCTION

The novel coronavirus disease 2019 (COVID-19) has become an emerging global health threat as a result of the rapid transmission across international borders.[1–3] This virus was first identified in Wuhan, Hubei Province, China, in late 2019 and now spread to the whole world.[4,5] There have been 101,053,721 confirmed cases and 2,182,867 deaths worldwide, mainly due to human-to-human transmission as of 29th January 2021.[6] Moreover, this zoonotic virus has caused most deaths compared to other CoVs.[7] Also, the reported deaths have crossed the deaths due to influenza, Zika, and Ebola. It is another headache that the pandemic continues due to improper policy interventions.[8] This pandemic has impacted the socioeconomic sector and healthcare systems drastically. As per reports, the global economic growth is estimated to be –3.0% in 2020 as compared to the previous year (+2.9%) in 2019.[9]

The accurate prediction of the disease to provide high-quality healthcare service management is a situational demand and challenging endeavor.[10] Artificial neural networks are relatively novel techniques and have been extensively exploited in numerous fields, such as agriculture,[11] environmental science,[12,13] artificial intelligence (AI),[14] epidemiology, and public health.[15,16] Artificial neural networks may provide valuable suggestions for chronic diseases.[17] Badnjevic et al.[18] have used ANNs to classify asthma and suggested the appropriateness of techniques for asthma detection. Moreover, Hasan in 2020[19] has predicted the upcoming COVID-19 situation by using EEMD-ANN Hybrid Model. Mollalo et al.[20] have also predicted the rate of COVID-19 incidence in the United States with the widely used topologies of ANNs. Some studies have also been reported for the modeling of the recent pandemic COVID-19 worldwide in different countries by using artificial intelligence and machine-learning algorithms. Although AI has the potential to assist our fight against COVID-19 and other future, the majority of the AI systems are currently in their preliminary stage and will take more time to obtain a satisfactory and visible result. However, there is no doubt in the fact that data is central in deciding the utility of AI-based models for managing future outbreaks.[21] The rapidly growing coronavirus-related datasets can be mined using AI to fuel our fight against this pandemic and can be utilized for developing vaccines, drug discovery, and public awareness. AI and AI-based models can help to rapidly predict and forecast the outcome and impact of outbreaks.[22]

AI-based models and techniques, including support vector regression, quantile random forest, *k*-nearest neighbors, cubist regression, and Bayesian regression neural network, were employed for forecasting COVID-19 incidences in the USA and Brazilian states. Based on the high accuracy (70%) and efficiency to predict COVID-19 cases in a time frame of only 6 days before, the tested models are likely to be suggested as potential forecasting tools to combat the adverse impacts of the current pandemic.[23] Abdulaal et al.[24] proposed an ANN for analyzing a number of patient-related features, including comorbidities, anthropometrics, demographics, smoking history, and lifestyle factors. It also predicted the mortality risk scoring of patients admitted to hospitals in the United Kingdom during the current pandemic. The model enabled predicting patient-specific mortality with accuracy, specificity, and sensitivity of 86.25%, 85.94%, and 87.50%, respectively.

As an alternative to replacing epidemiological models, Hu et al.[25] developed an AI-based approach to estimate the transmission dynamics, duration, size, and ending time of COVID-19 in China. Results revealed that the AI-inspired model showed good accuracy to forecast the COVID-19 trajectory. In the case of data reliability and lack of the second transmission, the employed model correctly indicated the transmission dynamics across the cities and provinces in China. Therefore, AI-based methods are regarded as useful tools in assisting policymaking and planning in healthcare. Our present study proposed an artificial intelligence (AI) approach using an artificial neural network to forecast the COVID-19 cases in Odisha for the upcoming days.

6.2 MATERIAL AND METHODS

6.2.1 ARTIFICIAL NEURAL NETWORK (ANN)

Figure 6.1 shows the l-m-m-m-n (l: input neurons, m: hidden layers, and n: output neurons) architecture of a back-propagation neural network. The input layer collects information from an external source and delivers it to the network for processing. The hidden layer is responsible for receiving raw information from the input layer and processing; this process is obtained by the output layer from the network and shows the results to an external

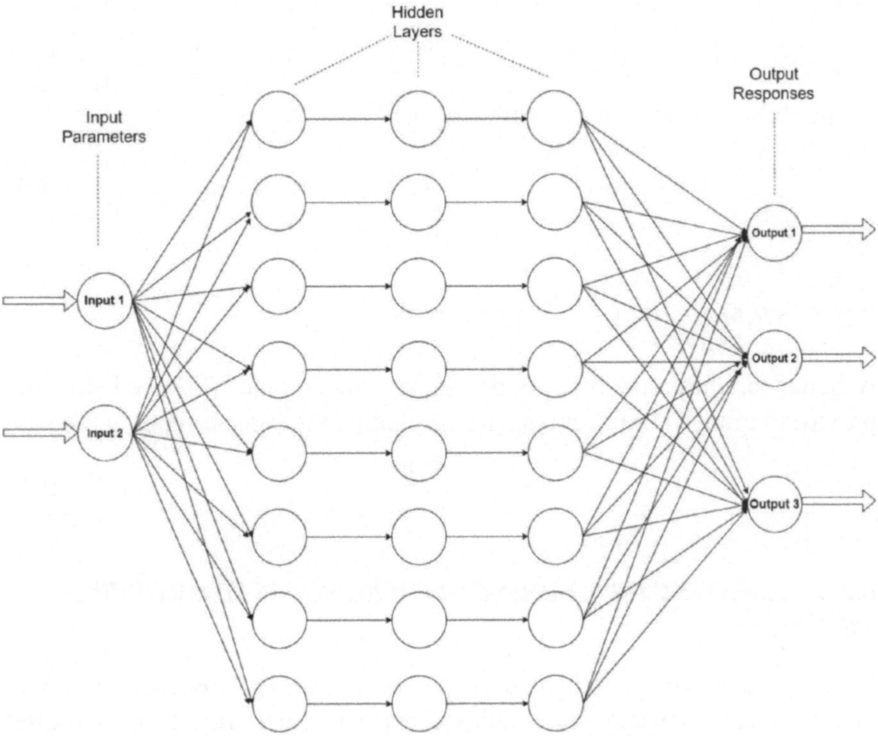

FIGURE 6.1 ANN architecture for the back-propagation COVID-19 model.

receptor. The input signals are modified by interconnection weight, recognized as a weight factor w_{ij}. A sigmoidal transfer function then further modifies the sum of modified signals (f). Likewise, the hidden layer's output signals are amplified by the interconnection weight w_{jk} of the kth node of the output layer to the jth node of the hidden layer. Finally, a pure linear transfer (f) function modifies the sum of the modified signal, and the output is received at the output layer.

Let $I_p = (I_{p1}, I_{p2}, \ldots\ldots I_{pl})$, $p = 1,2\ldots N$ be the pth pattern among N input patterns. W_{ji} and W_{kj} denote the connection weights between the ith input neuron to the jth hidden neuron, and the jth hidden neuron to the kth output neuron, respectively.

The output from a neuron in the input layer is

$$O_{pi} = I_{pi}, i = 1,2,\ldots.l \tag{6.1}$$

Output from a neuron in the hidden layer is

$$O_{pj} = f\left(NET_{pj}\right) = f\left(\sum_{i=0}^{l} W_{ji} O_{pi}\right), j = 1,2,......m \quad (6.2)$$

Output from a neuron in the output layer is

$$O_{pk} = f\left(NET_{pk}\right) = f\left(\sum_{i=0}^{l} W_{kj} O_{pj}\right), k = 1,2,......n \quad (6.3)$$

6.2.2 *SIGMOID TRANSFER FUNCTION*

A bounded, differentiable, nondecreasing, monotonic, *S*-shaped function provides a graded nonlinear response, including the logistic sigmoid function

$$f(x) = \frac{1}{1+e^{-x}} \quad (6.4)$$

6.2.3 *LEARNING OR TRAINING IN BACK-PROPAGATION NEURAL NETWORK*

In the present case, batch mode supervised learning was applied, where interconnection weights were adjusted using the delta rule algorithm after transferring the entire training sample to the network. During training, a comparison was made between the predicted and desired outputs, and the mean square error was calculated. If the mean square error (MSE) is higher than the prescribed limit, it is back-propagated from output to input, and weights are further modified till the error is brought within the prescribed limiting value.

Mean square error, E_p for pattern p is defined as

$$E_p = \sum_{i=1}^{n} \frac{1}{2}\left(D_{pi} - O_{pi}\right)^2 \quad (6.5)$$

Dpi and Opi are the target and output, and Opi is the computed output for the *i*th pattern.

At any time *t*, weight change is given by

$$\Delta W(t) = -\eta E_p(t) + \alpha \times \Delta W(t-1) \quad (6.6)$$

$$\eta = learning\ rate\ i.e\ 0 < \eta < 1$$

$$\alpha = momentum\ coefficient\ i.e\ 0 < \alpha < 1$$

6.3 RESULTS AND DISCUSSION

6.3.1 ARTIFICIAL NEURAL NETWORK (ANN) FOR PREDICTING COVID CASES

With an increase in COVID-19 cases, it is indispensable to comprehend the nature and upcoming trend of the situation. Various studies have been executed to comprehend and forecast the COVID-19 cases all around the world. For example, Shah et al.[26] applied a mathematical model SIR for the accurate prediction of the spread, recoveries, and deaths of COVID-19-infected individuals in Pakistan. Significant pandemic parameters-related information was scrutinized continuously and updated in this model to forecast the COVID-19 ending time. COVID-19 and SARS were found to be genetically identical, but their causes are entirely dissimilar. In contrast to SARS, COVID-19 appears less deadly but likely to be more transmittable and infectious.[27] For gauging the COVID-19 positive cases in India, Arora et al.[28] utilized deep learning-based models, including convolutional long-short term memory (LSTM), Bi-directional, and deep LSTM. The proposed LSTM model presented good accuracy for short-term forecasting with an error of < 3% and < 8% for predicting daily and weekly cases. Strict preventive measures were recommended in reducing the spread in all zones. These models could be used for predicting COVID-19 cases at the state level or even in other countries.

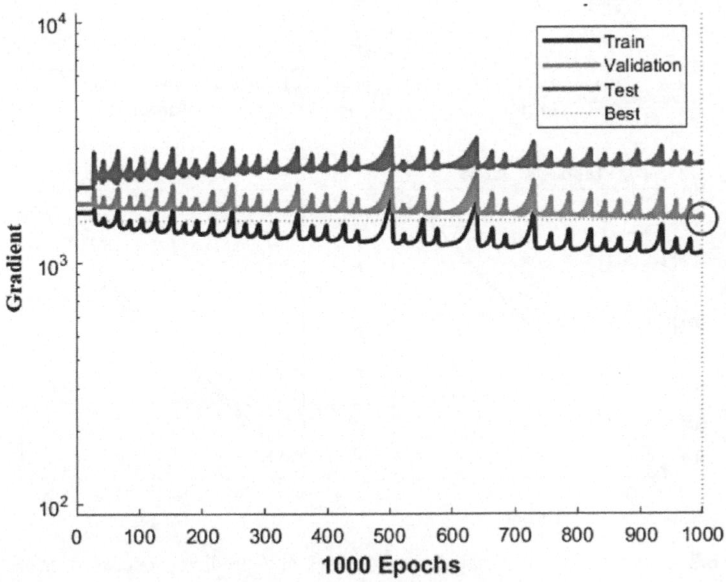

FIGURE 6.2 Convergence curve for COVID-19 deceased cases.

Patients' information-based algorithm was utilized to estimate the number of COVID-19-associated deaths in China. The death rate was speculated to fluctuate according to different temperatures and climates.[29] Ceylan[30] predicted the COVID-19 positive cases in France, Spain, and Italy using ARIMA models. The analysis and prediction of COVID-19 in Italy, China, and France have also been carried out.[31] In this predictive analysis, they estimated the total ventilation units' requirement and categorized the population as vulnerable, infected, recovered, and dead. In the present study, we have proposed artificial intelligence (AI) approach using an ANN to predict the COVID-19 cases in Odisha in the upcoming days. In the study, we have considered the number of days, the temperature at day and night time, and the humidity of the region as input parameters. But the cumulative number of deceased cases, confirmed cases, and active cases for COVID-19 in Odisha state are considered output responses for the study.

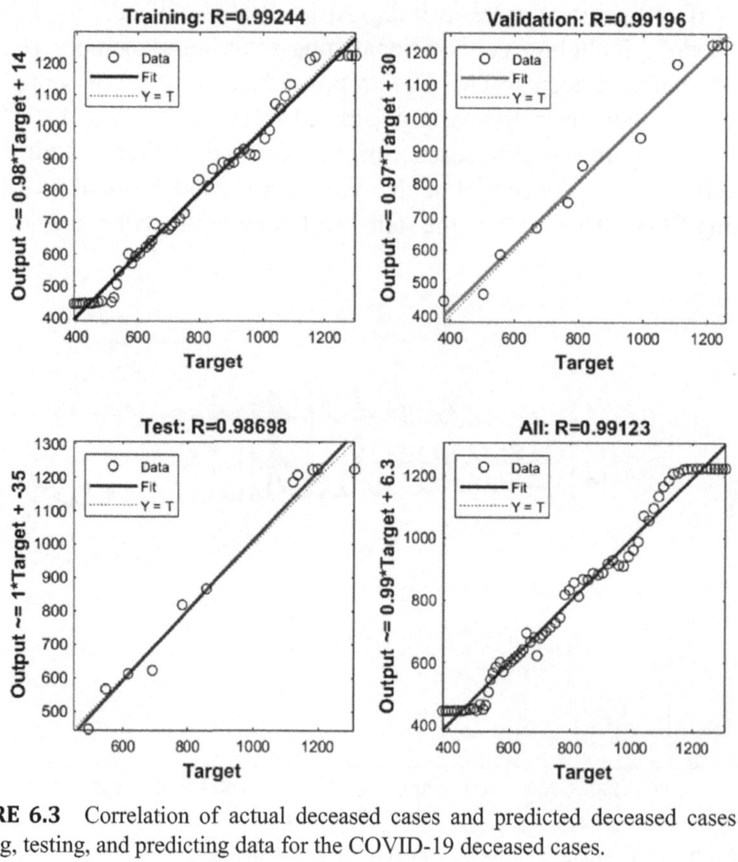

FIGURE 6.3 Correlation of actual deceased cases and predicted deceased cases during training, testing, and predicting data for the COVID-19 deceased cases.

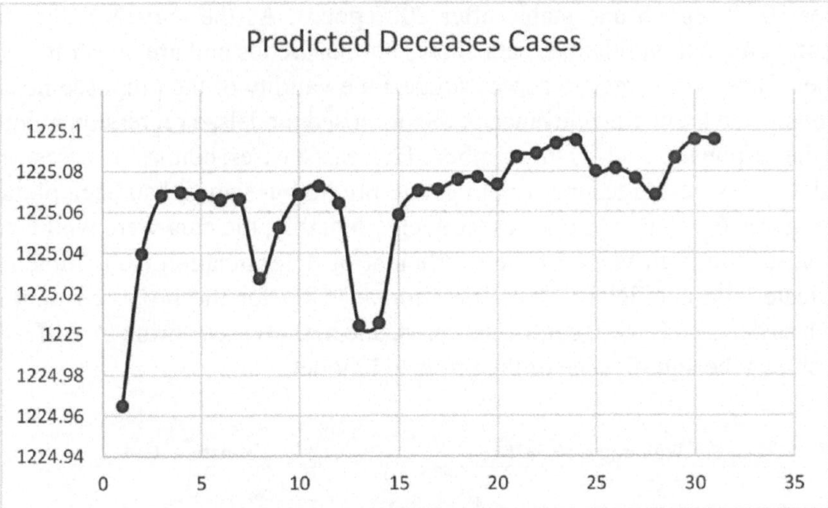

FIGURE 6.4 Predicted deceased cases for the upcoming days.

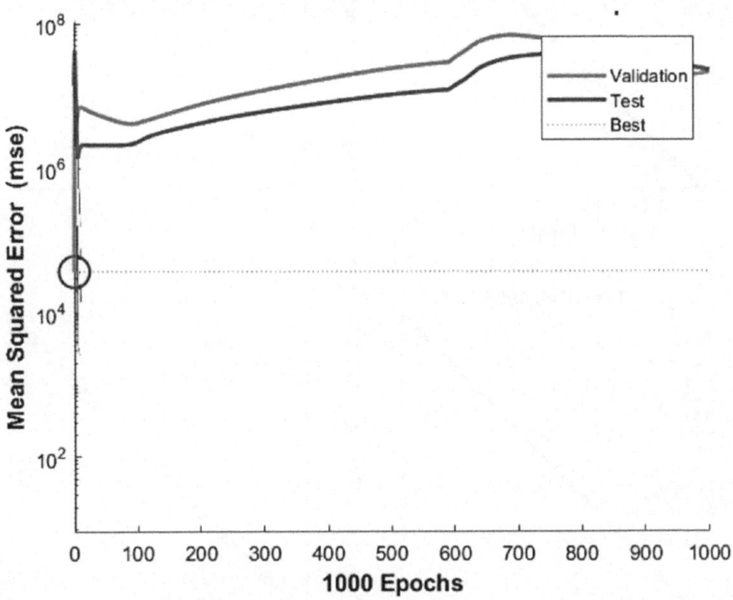

FIGURE 6.5 Convergence curve for COVID-19 confirmed cases.

To train the proposed network properly, 75% of data were utilized for training, and 25% of data were used for network testing, and no data was used twice. Figures 6.2, 6.5, and 6.8 show that the network was getting converged after 10–20 epoch and stable after 200 epochs. As the curve's nature for testing data and validating, data shows similar trends and are closer to each other. These convergence curves suggest the validity of the proposed neural network. To know the relevance of the proposed models, a regression plot for all the responses, such as the number of deceased cases, confirmed cases, and active cases during testing, training, and prediction stages, had been plotted in Figures 6.3, 6.6, and 6.9, respectively. Notably, the data were well fitted because of a high value of the coefficient of determination (R^2). Table 6.1 indicates the coefficient of the determinant (R^2) for the proposed outputs such as deceased cases, confirmed cases, and active cases during COVID-19. R^2 shows the significance of the projected model.

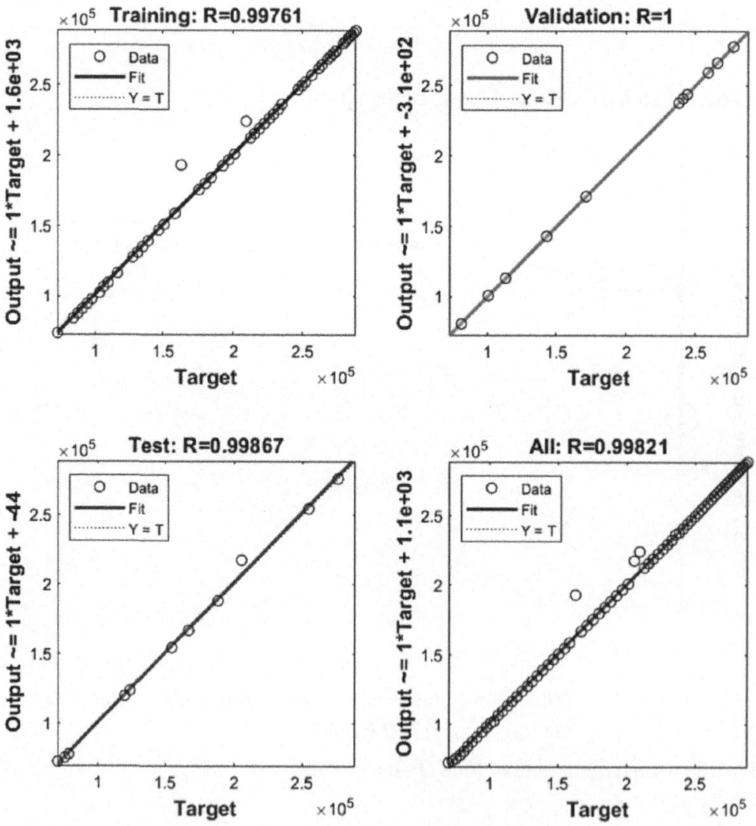

FIGURE 6.6 Correlation of actual confirmed cases and predicted confirmed cases during training, testing, and predicting data for the COVID-19 deceased cases.

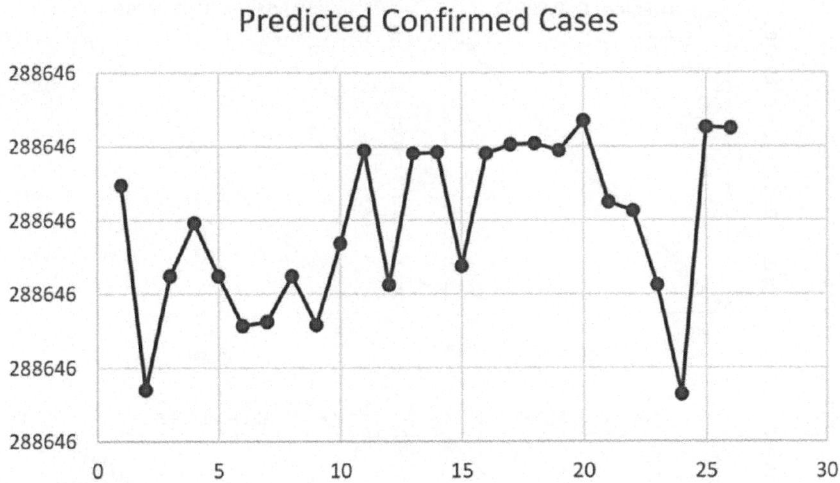

FIGURE 6.7 Predicted confirmed cases for the upcoming days.

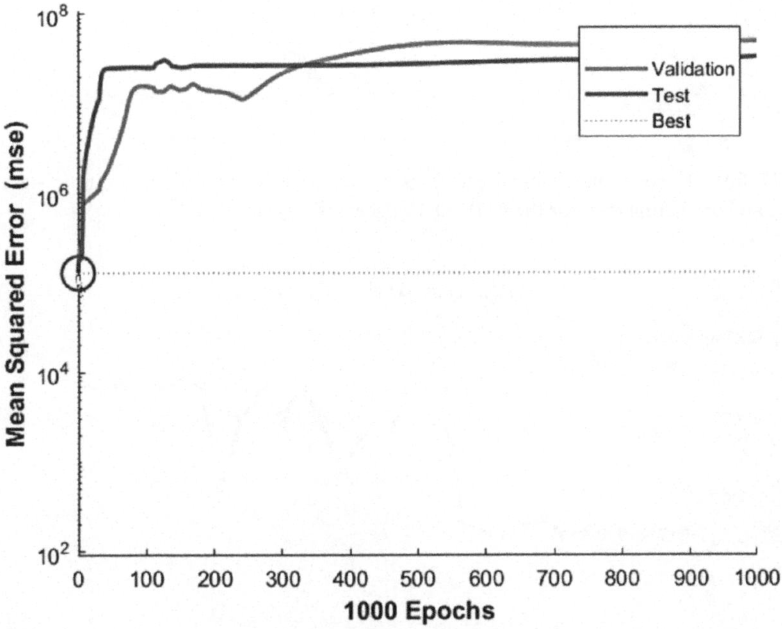

FIGURE 6.8 Convergence curve for COVID-19 active cases.

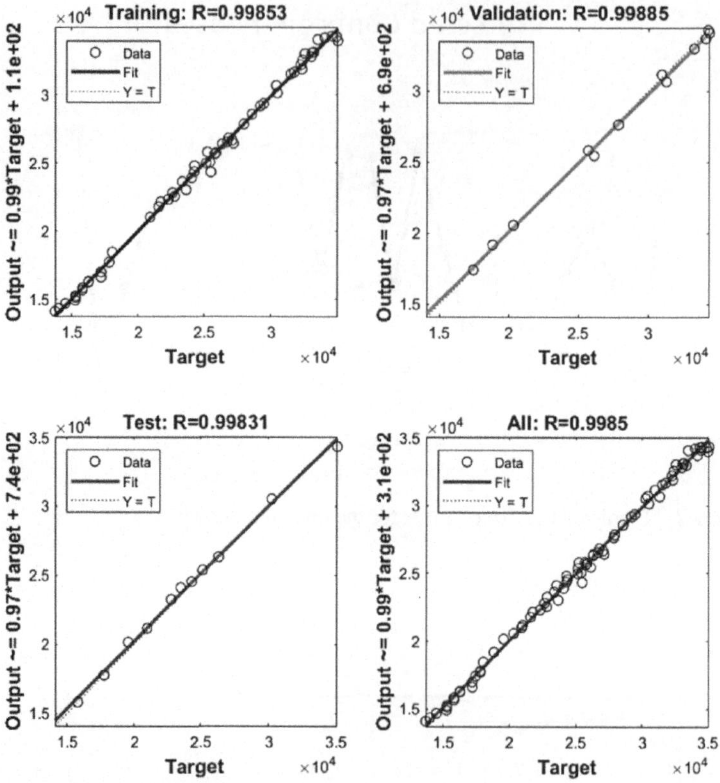

FIGURE 6.9 Correlation of actual active cases and active deceased cases during training, testing, and predicting data for the COVID-19 deceased cases.

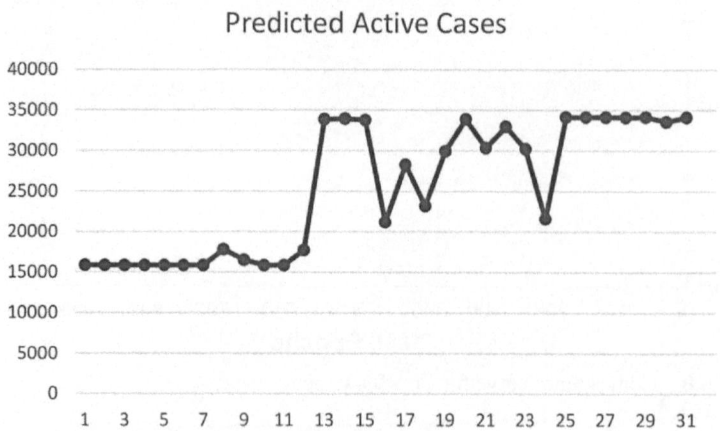

FIGURE 6.10 Predicted active cases for the upcoming days.

TABLE 6.1 The Coefficient of the Determinant (R^2) for the Proposed Outputs.

Coefficient of determinant (R2)											
Deceased cases				**Confirmed cases**				**Active cases**			
Train	**Test**	**Predict**	**All**	**Train**	**Test**	**Predict**	**All**	**Train**	**Test**	**Predict**	**All**
0.9924	0.987	0.992	0.9912	0.9976	0.9987	0.999≈1	0.9982	0.9985	0.9983	0.9989	0.9985

Train, training data; Test, testing data; Predict, validation/prediction data; All, overall model.

For the adequacy of the developed model, mean relative error was calculated between testing and predicted data obtained from the model. The mean relative errors of 0.0986, 0.0756, and 0.096 for a number of deceased cases, confirmed cases, and active cases, respectively, show the robustness in predicting the COVID-19 responses using the ANN by adapting the feed-forward back-propagation methodology. The proposed models for the network are found to be adequate and relevant in predicting the COVID-19 cases. Figures 6.4, 6.7, and 6.10 represent the upcoming trend for the number of deceased cases, confirmed cases, and active cases in the coming days. The results suggest an increase in COVID-19 cases.

6.4 CONCLUSION

The uncertainty in the growth rate of COVID-19 cases throws a spanner in its accurate prediction for the healthcare service management. The issue of formulating a model for the accurate prediction of the growth rate of COVID-19 infection further gets deranged owing to its situational and demographic dependence. Hence, any single model cannot be reliably applicable universally for such a prediction. A reliable prediction trend may pave the way for formulating a new strategy for restricting the proliferation of the viral infection. In our present study, we have examined an AI approach using an ANN for predicting the incidence of COVID-19 in Odisha in the upcoming days. We have considered the number of days, the temperature at day time and night time, and humidity of the region as input parameters. But the cumulative numbers of deceased cases, confirmed cases, and active cases for COVID-19 in Odisha state are considered output responses. For the adequacy of the developed model, the mean relative error was calculated between testing and predicted data obtained from the model. The mean relative errors of 0.0986, 0.0756, and 0.096 for the number of deceased cases, confirmed cases, and active cases, respectively, show the robustness in predicting the COVID-19 responses by adapting the feed-forward back propagation methodology. The proposed models for the network are adequate and relevant in predicting COVID-19 cases. Therefore, we hope the study may provide useful information for monitoring the COVID-19 outbreak.

CONFLICTS OF INTEREST

There are no conflicts to declare.

FUNDING

No funding received.

ORCID

Sharun Khan: https://orcid.org/0000-0003-1040-3746
Muhammad Bilal: https://orcid.org/0000-0001-5388-3183
Kuldeep Dhama: http://orcid.org/0000-0001-7469-4752
Om Prakash Jena: https://orcid.org/0000-0002-3088-974X

KEYWORDS

- **COVID-19**
- **predictive analysis**
- **artificial intelligence**
- **artificial neural network**
- **Odisha**

REFERENCES

1. Dhama, K.; Khan, S.; Tiwari, R.; Sircar, S.; Bhat, S.; Malik, Y. S.; Singh, K. P.; Chaicumpa, W.; Bonilla-Aldana, D. K.; Rodriguez-Morales, A. J. Coronavirus Disease 2019-COVID-19. *Clin. Microbiol. Rev.* **2020,** *33* (4), e00028-20.
2. Mohapatra, R. K.; Pintilie, L.; Kandi, V.; Sarangi, A. K.; Das, D.; Sahu, R.; Perekhoda, L. The Recent Challenges of Highly Contagious COVID-19; Causing Respiratory Infections: Symptoms, Diagnosis, Transmission, Possible Vaccines, Animal Models and Immunotherapy. *Chem. Biol. Drug Des.* **2020,** *96*, 1187–1208.

3. Mohapatra, R. K.; Das, P. K.; Kandi, V. Challenges in Controlling COVID-19 in Migrants in Odisha, India. *Diabetes Metab. Syndr. Clin. Res. Rev.* **2020,** *14*,1593–1594.
4. Fauci, A. S.; Lane, H. C.; Redfield, R. R. Covid-19—Navigating the Uncharted. *N. Engl. J. Med.* **2020,** *382*, 1268–1269.
5. Mohapatra, R. K.; Perekhoda, L.; Azam, M.; Suleiman, M.; Sarangi, A. K.; Semenets, A.; Pintilie, L.; Al-Resayes, S. I. Computational Investigations of Three Main Drugs and their Comparison with Synthesized Compounds as Potent Inhibitors of SARS-CoV-2 Main Protease (Mpro): DFT, QSAR, Molecular Docking, and in Silico Toxicity Analysis. *J. King Saud Univ. Sci.* **2021,** *33*, 101315.
6. World Health Organization. WHO Coronavirus Disease (COVID-19) Dashboard. Data Last Updated: Jan 29, 2021.
7. National Institutes of Health. COVID-19, MERS & SARS [Online]. https://www.niaid.nih.gov/diseases-conditions/covid-19.
8. United Nations. Everyone Included: Social Impact of COVID-19 [Online]. https://www.un.org/development/desa/dspd/everyone-included-covid-19.html.
9. International Monetary Fund (IMF). World Economic Outlook Chapter 1: The Great Lockdown [Online]. https://www.imf.org/en/Publications.
10. Chang, A. C. Artificial Intelligence and COVID-19: Present State and Future Vision. *Intell. Based Med.* **2020,** *3-4*, 100012.
11. Abdipour, M.; Younessi-Hmazekhanlu, M.; Ramazani, S. H. R. Artificial Neural Networks and Multiple Linear Regression as Potential Methods for Modeling Seed Yield of Sa_ower (Carthamus tinctorius L.). *Ind. Crop. Prod.* **2019,** *127*, 185–194.
12. Keshavarzi, A.; Sarmadian, F.; Sadeghnejad, M.; Pezeshki, P. Developing Pedotransfer Functions for Estimating Some Soil Properties Using Artificial Neural Network and Multivariate Regression Approaches. *ProEnviron. Promediu* **2010,** *3*, 322–330.
13. Marohasy, J.; Abbot, J. Assessing the Quality of Eight Different Maximum Temperature Time Series as Inputs When Using Artificial Neural Networks to Forecast Monthly Rainfall at Cape Otway, Australia. *Atmos. Res.* **2015,** *166*, 141–149.
14. Kang, B. H.; Bai, Q. In *AI 2016: Advances in Artificial Intelligence*, Proceedings of the 29th Australasian Joint Conference, Hobart, TAS, Australia, Dec 5–8, 2016; Springer: Berlin/Heidelberg, Germany, 2016; vol 9992.
15. Kiang, R.; Adimi, F.; Soika, V.; Nigro, J.; Singhasivanon, P.; Sirichaisinthop, J.; Leemingsawat, S.; Apiwathnasorn, C.; Looareesuwan, S. Meteorological, Environmental Remote Sensing and Neural Network Analysis of the Epidemiology of Malaria Transmission in Thailand. *Geospat. Health* **2006,** *1*, 71–84.
16. Mollalo, A.; Sadeghian, A.; Israel, G. D.; Rashidi, P.; Sofizadeh, A.; Glass, G. E. Machine Learning Approaches in GIS-Based Ecological Modeling of the Sand Fly Phlebotomus Papatasi, A Vector of Zoonotic Cutaneous Leishmaniasis in Golestan Province, Iran. *Acta Trop.* **2018,** *188*, 187–194.
17. Reddy, R.; Imler, T. D. Artificial Neural Networks are Highly Predictive for Hepatocellular Carcinoma in Patients with Cirrhosis. *Gastroenterology* **2017,** *152*, S1193.
18. Badnjevic, A.; Gurbeta, L.; Cifrek, M.; Marjanovic, D. In *Classification of Asthma Using Artificial Neural Network*, MIPRO, Proceedings of the International Convention, Proceedings of the 2016 39th International Convention on Information and Communication Technology, Electronics and Microelectronics (MIPRO), Opatija, Croatia, May 30–June 3, 2016; IEEE: Piscataway, NJ, USA, 2016; pp 387–390.

19. Hasan, N. A Methodological Approach for Predicting COVID-19 Epidemic Using EEMD-ANN Hybrid Model. *IoT* **2020,** *11*, 100228.
20. Mollalo, A.; Rivera, K. M.; Vahedi, B. Artificial Neural Network Modeling of Novel Coronavirus (COVID-19) Incidence Rates across the Continental United States. *Int. J. Environ. Res. Public Health* **2020,** *17*, 4204.
21. Naudé, W. Artificial Intelligence vs COVID-19: Limitations, Constraints and Pitfalls. *AI Soc.* **2020,** *35* (3), 761–765. DOI: 10.1007/s00146-020-00978-0. Epub ahead of print. PMID: 32346223; PMCID: PMC7186767.
22. Ahuja, A. S.; Reddy, V. P.; Marques, O. Artificial Intelligence and COVID-19: A Multidisciplinary Approach. *Integr. Med. Res.* **2020,** *9* (3), 100434. DOI: 10.1016/j.imr.2020.100434. Epub 2020 May 27. PMID: 32632356; PMCID: PMC7255319.
23. da Silva, R. G.; Ribeiro, M. H. D. M.; Mariani, V. C.; dos Santos Coelho L. Forecasting Brazilian and American COVID-19 Cases Based on Artificial Intelligence Cupled with Climatic Exogenous Variables. *Chaos Solitons Fractals* **2020,** *139*, 110027.
24. Abdulaal, A.; Patel, A.; Charani, E.; Denny, S.; Mughal, N.; Moore, L. Prognostic Modeling of COVID-19 Using Artificial Intelligence in the United Kingdom: Model Development and Validation. *J. Med. Int. Res.* **2020,** *22* (8), e20259.
25. Hu, Z.; Ge, Q.; Jin, L.; Xiong, M. Artificial Intelligence Forecasting of Covid-19 in China. *arXiv* 2020; *preprint arXiv: 2002.07112.*
26. Shah, S. T. A.; Mansoor, M.; Mirza, A. F.; Dilshad, M.; Khan, M. I.; Farwa, R.; Khan, M. A.; Bilal, M.; Iqbal, H. M. N. Predicting COVID-19 Spread in Pakistan Using the SIR Model. *J. Pure Appl. Microbiol.* **2020,** *14* (2), 1423–1430.
27. Huppert, A.; Katriel, G. Mathematical Modelling and Prediction in Infectious Disease Epidemiology. *Clin. Microb. Infect.* **2013,** *19* (11), 999–1005.
28. Arora, P.; Kumar, H.; Panigrahi, B. K. Prediction and Analysis of COVID-19 Positive Cases Using Deep Learning Models: A Descriptive Case Study of India. *Chaos Solitons Fractals* **2020,** *139*, 110017.
29. Wang, L.; Li, J.; Guo, S.; Xie, N.; Yao, L.; Cao, Y.; Day, S. W.; Howard, S. C.; Graff, J. C.; Gu, T.; Ji, J.; Gu, W.; Sun, D. Real-Time Estimation and Prediction of Mortality Caused by COVID-19 with Patient Information-Based Algorithm. *Sci. Total Environ.* **2020,** *727*, 138394.
30. Ceylan, Z. Estimation of COVID-19 Prevalence in Italy, Spain, and France. *Sci. Total Environ.* **2020,** *729*, 138817.
31. Fanelli, D.; Piazza, F. Analysis and Forecast of COVID-19 Spreading in China, Italy and France. *Chaos Solitons Fractals* **2020,** *134*, 109761.

CHAPTER 7

IoT in Healthcare Ecosystem

POONAM GUPTA[1], INDHRA OM PRABHA M.[1], and AHMED A. ELNGAR[2]

[1]*G H Raisoni College of Engineering and Management, Pune, Maharashtra, India*

[2]*Faculty of Computers and Artificial Intelligence, Beni-Suef University, Beni-Suef City, Egypt*

ABSTRACT

In recent years, the IoT has revolutionized the technology landscape, driving explosive growth in areas such as automated manufacturing, asset management, and wearable consumer health products. The presence of IoT can be seen in every field. Healthcare has undergone a major transformation since IoT entered the medical field. Personal healthcare is becoming a reality as IoT offers solutions on multiple dimensions. This chapter provides a comprehensive overview of IoT in the healthcare ecosystem, discussing various applications of IoT in healthcare, architectural models, challenges facing IoT in healthcare, security practices and issues, and Learn about the future of IoT domain. The first section covers medical IoT applications, patient applications such as remote health monitoring and critical care monitoring, and hospital IoT applications such as staffing, record time reduction, and real-time location of medical equipment. I will explain. We then discuss how data collected from patient- and hospital-centric applications can help drive other areas such as health insurance. He then discusses his IoT support for the pharmaceutical industry to limit counterfeit medicines. Next, we will discuss the implementation design of healthcare IoT. In addition to traditional cloud services, new services such as fog and edge computing have increased significantly in recent years. Fog computing and edge computing

Global Digital Transformation and the Covid-19 Pandemic.
Rajdeep Chowdhury, K. Martin Sagayam, Neyara Radwan, Ahmad A. Elngar (Eds.)

are viewed as intelligent and flexible architectures. The next section discusses the design of the architecture and the advantages and challenges of the two computational models. The next section details how the two applications were actually implemented. The next section describes the challenges facing IoT in healthcare. In general, challenges can be categorized into technical challenges, human-centric challenges such as the acceptance of his IoT domain in healthcare, and finally security bottlenecks. The data generated and maintained by IoT platforms will serve as a goldmine for various stakeholders, such as patients and their helpers, as well as medical professionals and insurance companies for future research and development in the medical field. Security aspects of how domains handle sensitive patient data are therefore more important. The following sections provide insight into the security issues, cyber threats, and attacks facing her IoT and defenses in healthcare. Additionally, this chapter looks at the role of IoT in the fight against the novel coronavirus that caused the unprecedented global pandemic. Finally, I will discuss the future of IoT. With the advent of 5G and the proliferation of artificial intelligence, various facets of IoT are being debated, which are expected to grow exponentially.

7.1 INTRODUCTION

As the IoT permeates healthcare, the healthcare ecosystem is undergoing a paradigm shift. However, the fact that the IoT leverages the existing ubiquitous presence of connected objects and networked resources within the healthcare platform facilitates the transition. The IoT helps explore new dimensions of patient care through remote care, faster disease diagnosis, aggressive treatment, and follow-up.[1] Also, with the help of the IoT, he strengthened the bond between the patient and the doctor as the relationship between the two parties became more efficient and easier.

Currently, the main operational bottlenecks facing healthcare systems are understaffed hospitals, rising costs of advanced medical services, and physicians who do not want to move to rural areas.[2,3] In addition, the growing aging population in need of remote care is of increasing concern in many countries. Another new challenge is the new COVID-19 disease, which puts additional strain on the already tense healthcare sector. The IoT revolution, achieved through the fusion of technology and innovation by scientific researchers, provides solutions to these medical needs. Building IoT-enabled healthcare systems has proven feasible for most clinical providers, as IoT leverages the healthcare infrastructure and networks already in place.

Doctor–patient ratios in most countries are consistently below the WHO-prescribed value of 1:1000.[4,5] In India, for example, there is one doctor for every 1445 inhabitants, so intelligent healthcare can help in this case. Telemedicine reduces patient visits to hospitals, disease diagnosis using smart devices without physician intervention,[6] postdischarge testing, and remote patient monitoring.[1,7] All of this is made possible today with modern healthcare services in the IoT revolution without the need for a doctor.

The healthcare industry involves high costs and high consulting fees that make up the management and delivery of facilities for onsite services. Access to some sophisticated devices and diagnostic procedures is still expensive. Health insurance premiums have skyrocketed in recent years, reflecting high medical costs. With the advent of the IoT in the healthcare sector, the cost of services can be reduced and made feasible.

Moreover, the biggest challenge in developing countries is the health care of rural populations.[2,3] Healthcare is a right of every citizen, and equal healthcare is inevitable wherever patients are. However, the current situation lacks quality infrastructure, lack of access to basic medicines, and lack of local qualified health professionals. In a country like India, where 65% of the population lives outside villages, the lack of reach to health facilities in rural areas is seen as a major failure. IoT has proven to be a promising technology to fill this gap. The IoT provides cost-effective remote health monitoring and facilitates the availability of health devices. This makes healthcare affordable and accessible, as described in article[3] IoT-based Smart Edge for Global Health: Remote Monitoring with Severity Detection and Alerting increase.

Today, the aging of the population is one of the biggest problems facing the healthcare industry in many countries. The elderly population is growing rapidly worldwide and seems to be an important part of society. The population aged 65 and over is projected to double by 2050 with an increase in life expectancy. This challenge must be met by all countries, regardless of their socioeconomic status. The IoT addresses this challenge and provides the best, cost-effective solution for improving the quality of life for older people. Some of the major problems faced by the elderly are social isolation, unplanned falls,[8] and medication compliance.[9–11]

Accidental falls are the second leading cause of death in the world. Adults over the age of 65 suffer from the most deadly falls, or falls that are life-threatening. With an IoT-based immutable fall detection system,[8] this paper proposes a solution using his IoT technology that integrates accelerometers,

GPS, and Wi-Fi into devices such as smartphones, Arduino, and Raspberry Pi boards. Simple designs are created using the client–server architecture. The data of continuous movement is sent to the server or cloud where the pattern is examined. And if anomalies are detected in the data, alerts are sent to connected medical facilities immediately. Taking this a step further, IoT can monitor and report whether patients are taking their medications as directed. A paper on medication adherence with real-time activity detection in IoT[9–11] describes how to track and monitor medication adherence. Loneliness can also be combated with the help of IoT-based robotics. It is friendly and can support the elderly in their daily activities, respond to commands, and act as companions.

This chapter is organized as follows: Section 2 describes the various IoT applications in the healthcare industry. Section 3 describes pioneering implementations of the IoT in healthcare. Then, Section 4 deciphers the architecture of some IoT models. Section 5 analyzes the challenges facing the IoT in healthcare. Section 6 discusses security issues and defense mechanisms. Section 7 details the role of the IoT in the fight against the global COVID-19 pandemic. Finally, Section 8 looks at the advent of 5G and the future of IoT through the breakout of synergies between artificial intelligence and IoT-generated data. This is followed by a conclusion to discuss the future face of the IoT.

7.2 APPLICATIONS OF IOT IN HEALTHCARE

IoT has a gamut of applications in the healthcare domain; this section classifies the applications into two broad categories patient-centric and hospital-centric. The patient-centric applications deal with remote patient monitoring and care, pathology detection, emergency patient care, and body wearable devices. The hospital-centric applications deal with the efficient use of hospital resources and the deployment of healthcare professionals. IoT also helps the pharma and medical insurance fields which also come under the umbrella of healthcare.

7.2.1 PATIENT-CENTRIC IOT

Rising chronic diseases and an aging population are increasing the demand for healthcare workers and hospitals. These challenges are met with advancements in the technologies like IoT, Artificial intelligence, and data analytics and employing them to the benefit of healthcare.

7.2.1.1 REMOTE PATIENT CARE

Remote patient care is a part of telehealth, where the patients are monitored remotely from their residences without the need to visit the hospital.[3] It is a widely preferred model for the elderly population and for people whose mobility is restricted.

- **Elderly people care**
  - Invariant fall detection

    Elderly people care is a pressing need, with more elderly people staying alone at home. Using a body sensor, like a smartphone or Raspberry Pi which has to be kept in the pocket of the patient records the movement of the patient. The accelerometer data is continuously read and sent to the server in the cloud where a machine learning algorithm is applied to the data and predicts any fall or abnormal motion. In Ref. [8] an IoT-based device-type invariant fall detection system is given.

- **People whose mobility is restricted**

  Quadriplegia patients are patients whose limbs are not functional. For such patients, IoT can help in multiple ways. The Healthcare Robotics Lab at Georgia Tech designed a robotically controlled bed called an Auto bed which can be controlled by a web-based interface.[12] So, such beds can be controlled remotely by the web application. Similarly, Smart Wheelchairs[13] using IoT is an innovative model that makes life easier.

- **Heart disease prediction**

  Patients with premedical conditions like high blood pressure can be kept under continuous monitoring.[6] The bodywear like smartwatch and ECG that are fixed in the patient's body reads the data and sends it to the cloud seamlessly. In case of any abnormalities, warning messages are sent to the patient and the immediate caretaker so that they can act immediately.

- **People with a chronic condition**
There are cases where the patients are not required to be hospitalized yet are under constant threat due to chronic conditions. For such patients, IoT solutions come as a boom. For instance, the IoT-enabled HVAC can be installed in the home or the living room of a patient who suffers from respiratory diseases like COPD for whom ventilation and air quality are crucial. The sensors in the HVAC read the pollution level of the air constantly and the smart HVAC device will adjust it back to the desired level.[14]

7.2.1.2 PATHOLOGY AND FATAL VIRAL/BACTERIAL DISEASES

Electroencephalogram (EEG) sensors are placed at key points on the patient to monitor electrical signals. EEG data is monitored along with facial expressions, speech, movements, and gestures. By performing data analysis, the disease is predicted as described in the publication[15] Cognitive Smart Healthcare for Pathology Detection and Monitoring.

7.2.1.3 CRITICAL AND EMERGENCY PATIENT CARE

Time, availability, and accuracy of information play a critical role when it comes to critical and emergency patient care. In most cases, emergency physicians do not know the medical history, and information collected while the patient is being transported by ambulance is often inaccurate. In such cases, a central database holding patient data for IoT healthcare serves as a potential backup. There, the emergency doctor is well informed about the patient's file until the patient arrives and has a plan of action ready. Associated losses are reduced, and emergency care is improved.[16]

7.2.1.4 FOOD AND WORKOUT MONITORING

Apart from critical patient care monitoring, food and workout monitoring is gaining popularity among fitness-conscious people and sportspersons. The wearable fitness trackers which are IoT-based track the physical activity of the person.[17] The wristband or smartwatch can track the number of steps taken, distance covered, calories burnt, heart points, average speed, body temperature, heart rate, and sleep quality of a person.

7.2.1.5 AFFECTIVE COMPUTING

This is the process of recognizing human emotions with the help of computer systems and devices. It can be characterized as an interdisciplinary discipline that includes computer science, psychology, and cognitive science. IoT devices help capture the emotional state of the user. Devices include sensors, microphones, and cameras.[18]

Affective computing technology captures a user's emotional state and responds by performing certain predefined product/service functions, such as: For example, changing the quiz to suit the learner's mood or recommending a series of videos. For example, if a quadriplegic patient who is unable to move his limbs wants to make an emergency call, the camera uses her computing to capture facial expressions and recognize his voice. Perform the prescribed actions. Alternatively, with the help of artificial intelligence, actions are determined using predictive methods. In other words, effective computing can be thought of as a human–computer interaction in which a computer responds based on human facial expressions, stimuli, gestures, utterances, or postures. In addition, the temperature change of the hand recorded by the mouse can also be used as an input for affiliate computing. Effective computing, therefore, takes advantage of human physiological and emotional changes to predict what actions to take.

7.2.2 HOSPITAL-CENTRIC IOT APPLICATIONS

IoT devices and applications play a major role in optimizing hospital management and operations apart from patient care. The operations management in the hospital is essential to control the costs and improve the quality of service provided to the patients. The hospitals achieve this by employing IoT.

7.2.2.1 REAL-TIME LOCATION OF MEDICAL EQUIPMENT

Sensored IoT devices are used to track the real-time location of medical devices such as wheelchairs, defibrillators, nebulizers, oxygen pumps, and other surveillance devices.[19]

7.2.2.2 DEPLOYMENT OF MEDICAL STAFF

Medical staff deployment can be monitored with the help of IoT. The movement of the hospital staff and doctors can be kept track and in case of emergency, the nearest doctor to the location can be referred immediately thereby gaining on the time in doctor reaching the critical patient within the hospital premises.[19]

7.2.2.3 DRUGS MANAGEMENT

Management and governance of drugs is a major challenge in the medical sector. Apparently, there are high chances of counterfeit drugs replacing genuine medicines. And there are many drugs that are sensitive to parameters like temperature and humidity. These are managed by implementing IoT and blockchain technology.[20] For instance, some drugs that are temperature sensitive are kept in refrigerators or customized medical boxes where the temperature and humidity are maintained as recommended by the drug manufacturer.[21]

7.2.2.4 REDUCING THE CHARTING TIME

Patient charting is one of the important works of doctors which is also a considerably time-consuming activity. Patient charting keeps complete information of the patient on file, including demographics, diagnoses, medications, vital signs, allergies, lab/test results, treatment plans, immunization dates, progress notes, and more. IoT provides a solution for this by recording the voice commands of the physician. It makes the patient's data readily accessible for review, thereby saving 15 hours per week of doctor's time.

7.2.3 IOT BENEFITTING HEALTH INSURANCE COMPANIES

Insurance companies can benefit from data collected by the healthcare industry. You can use this data to perform underwriting tasks and claim assessments. Fraudulent claims can be easily inferred, as the person's medical history contains all the necessary data. In this way, data collected by IoT devices bring transparency between customers and insurers. Data transparency simplifies all insurance processes such as underwriting, pricing, billing, and claims.

7.2.4 PHARMACEUTICAL GOVERNANCE

The Pharmaceutical industry always faces threats posed by parallel counterfeit drug manufacturers. Apparently, the infusion of fraudulent drugs in place of legitimate drugs has increased in the past decade. It is a huge challenge to identify and curb this parallel counterfeit drug supply chain as their existence is completely invisible. IoT along with Blockchain has helped the pharma industry in addressing this challenge by enhancing the traceability of medicines in the pharma supply chain.[20] The drugs are tracked from the source till they reach the customers without any breach or intrusion.

7.3 IMPLEMENTATION METHODOLOGIES

The standard implementation of the IoT model involves the IoT devices/ sensors directly connected to the cloud and hence the data generated by the sensors would be directly sent to the cloud. The storage and processing of data were done in Cloud. However, with the growing rate of IoT and innumerable devices getting connected to the cloud, this model started to experience several gridlocks like the storage space, processing speed, and mainly the bandwidth.

This made way for the concept of Fog and Edge computing which took processing closer to the source where the data is created, thereby filtering only the required and already processed data to be sent to the cloud and eliminating the unwanted data at the Edge or Fog layer itself. By doing so, the processing speed is improved, storage space is optimized, and the bandwidth is saved.

7.3.1 FOG COMPUTING

Fog computing has only one centralized computing device that processes data from different endpoint IoT devices.[22,23] Edge computing, on the other hand, processes all data on multiple nodes. Fog computing ensures that computing power and data analytics are transferred to your local network or IoT gateway.

7.3.1.1 ARCHITECTURE

The architecture of the Fog computing is shown below

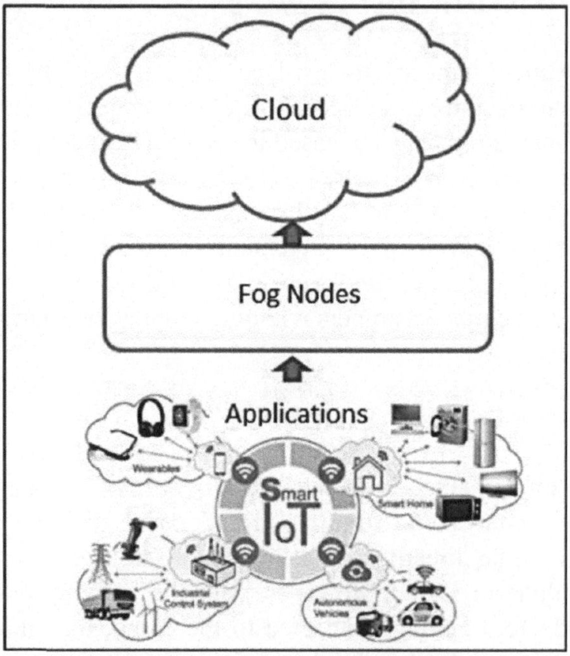

FIGURE 7.1 Standard implementation of fog computing model.

The fog computing structure includes 3 foremost layers:

- **Smart IoT devices/applications**
  These are the devices like sensors, and actuators from the source that generate data and transmit it to the Fog network and eventually to the cloud.
- **Fog nodes**
  Fog nodes collect data in real time from multiple sources and processes. It can handle multiple protocols. Fog nodes then run IoT-enabled applications for real-time analytics, often with millisecond response times.
- **Cloud**
  The cloud receives and compiles periodic data summaries sent by various fog nodes. It also performs data analysis to gather

business insights and sends new application rules to fog nodes based on those insights.

Fog computing is responsible for analyzing the most time-sensitive data transmitted by the Internet of Things. Basically, the fog node closest to the sensor looks for problems or signs and blocks them from sending signals to the actuators. Data that can only wait seconds or minutes is sent directly to the nebula node for analysis and action. Dates that are not time-sensitive are sent directly to the cloud for long-term storage or historical analysis. Fog nodes periodically send data to the cloud for storage. By effectively using local data and sending only selected data to the cloud, fog computing can save bandwidth.

In Ref. [23] the adaptive and fault-tolerant data processing of healthcare IoT based on fog computing, reliable data transmission, and fast data processing speed are regarded as key aspects of fog computing, and we will improve it. A framework has been developed to improve data transmission reliability and processing speed in fog computing, a Reduced Variable Neighborhood Search (RVNS)-based sensor data processing framework (REDPF) was developed.

7.3.1.2 ADVANTAGES

- Low latency—As the processing nodes are closer to the user the latency in fog computing is negligible.
- High Security—Security is handled by multiple nodes.
- No issues related to bandwidth as the data is not sent in bulk.

7.3.2 EDGE COMPUTING

Edge computing uses local computing and local edge devices to deliver intelligent services.[24,25] A local edge device can be a router, smartphone, PC, or ATM.

7.3.2.1 ARCHITECTURE

Below is the architectural diagram showing different layers of Edge computing. The two layers between the IoT devices and the cloud layer constitute the Edge components. The Edge nodes communicate with the Edge IoT Gateway. For instance, if a nuclear reactor sends the temperature every second and the cloud-based analytical application requires the consolidated data for a minute, the second's data is stored and processed in the edge node, and only the minute data is sent to the cloud.

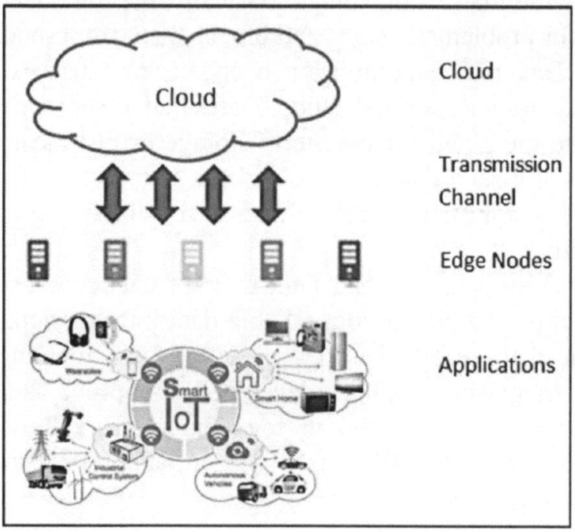

FIGURE 7.2 The standard implementation model of edge computing.

- **IoT nodes/applications**

  This includes all the devices and applications that generate data at regular intervals. The IoT devices in some cases act as Edge nodes as well.

- **Edge nodes**

  The Edge nodes can be separate nodes or can be the same IoT device, the edge nodes are responsible for data generation, collection, preprocessing, and analytics. They are connected to sensors, actuators, and smart devices. These edge nodes help to decentralize data processing and lower the dependence on the cloud.

- **Cloud**

  The cloud receives regular data from the gateway and performs data analysis on that data.

In Ref. [24], an Edge-based Architecture to Support Efficient Applications for Healthcare Industry 4.0 proposes a body-edge framework that includes edge computing with a 3-tier architecture (IoT Edge Device, Edge Gateway, Cloud Tier). Body Edge consists of a mobile client module and an IoT Edge gateway that supports multi-radio and multi-technology communication to collect data from various scenarios and process it locally. The proposed model has been shown to reduce traffic to the Internet and increase processing time.

7.3.2.2 ADVANTAGES

- Faster response times because data is processed locally
- Preprocess and filter the data
- Aggregate the data to send to the cloud
- Reduce the bandwidth used to transmit data to the cloud
- Low operating cost as the data to be managed by the cloud is reduced
- Lesser delays as the processing is split into multiple nodes
- Improved performance due to low latency
- Unlimited scalability
- Conserving network

7.3.2.3 *EMPOWERING EDGE COMPUTING*

Edge nodes can process data in real time. If we can add intelligence to the Edge node, the Edge device can perform data analysis on its side instead of relying on the cloud. For example, if deep learning architectures like convolutional neural networks can be added to Edge computing, data analysis and decision-making can be brought closer to the source, reducing latency and saving cost and bandwidth.

Real-time outcomes can be seen in ATMs which act as the edge device and transmit every transaction to the cloud. With the help of deep learning architecture embedded in the ATMs, it becomes possible for the ATMs cameras to detect if someone is accessing the ATM wearing cap or helmet. Instead of relying on the cloud for the processing which could take a delay in time, real-time processing can be done at the ATM itself.

If analytics can be brought to the edge devices, it becomes flexible to scale up and modify the analytics at the node level as compared to the cloud without affecting the seamless operations of the device.

7.4 IMPLEMENTATION MODELS

Two implementation models of IoT healthcare are discussed in detail in this section which provides an insight into the design from end to end. It explains the way data is collected from the sensors attached to the patient's body, transferred to the cloud, and finally to the repository where the exhaustive analysis of data happens and the results are predicted.

7.4.1 HEART DISEASES PREDICTION

IoT breakthrough in heart disease prediction is a niche in the medical landscape. Heart disease is the major contributor to medical deaths. Such heart diseases can be tracked and monitored by implementing the IoT.

In Ref. [26], the author proposed a new system using Modified Deep Convolution Neural Network (MDCNN). The system includes two sensors attached to the patient's body that collect data and send it to the MDCNN. The two sensor devices are smartwatches and heart monitors. They monitor the patient's blood pressure and electrocardiogram (ECG). The collected data is subjected to deep learning analysis. There are three steps through which the data will pass before the prediction is made. They are (1) preprocessing, (2) feature selection, and (3) classification using a modified deep convolutional neural network. The accuracy of this method is 98.2 times better than other deep learning and logistic regression methods.

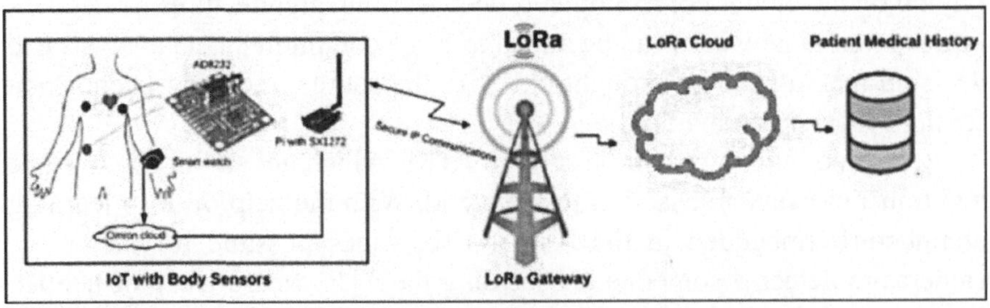

FIGURE 7.3 IoT framework of heart disease prediction model.

The IoT design of the framework is shown in Figure 7.3. The data is collected from the wearable devices that is worn by the patient, the smartwatch and the heart monitor board AD8232. These read the patient's resting blood pressure and electrocardiographic data. The smartwatch used is Omron HearGuid-bp8000m which can send data to the Omron cloud and thereby send alert messages to the concerned mobile numbers when the readings go abnormal. The data collected from the IoT sensor devices are captured and processed using Raspberry Pinboard. This data is then sent through a secured communication low-range (LoRa) Gateway and to the LoRa Cloud.

LoRA is a low-range spread spectrum modulated technique where packets can be sent to a range of a few kilometers. It is a low-power protocol. It requires low power and hence the battery can last a year long. The longevity of the battery is achieved by powering up and down the battery. The LoRA transmitter is powered up only when the packets are sent instead of keeping them always powered like the Wi-Fi radio. The low-range protocol is limited to transmitting only a few hundred bytes per transmission.

7.4.2 HEALTHCARE IOT-BASED AFFECTIVE STATE MINING USING DEEP CNN

Besides predicting the physiological conditions of people, IoT applications are also used to predict the psychological conditions of people. The IoT sensor devices record the horological observations with the help of electro-myography (EMG), electrocardiogram (ECG), and electro-dermal activity (EDA) medical sensors.[18]

- **Electro-dermal Activity**

  EDA refers to the property of continuously changing the electrical properties of the skin. Sweat glands are controlled by the sympa-thetic nervous system, and when aroused they begin to sweat from the skin. Therefore, the electrical properties of the skin (conductivity or resistance) can be used to measure people's emotional and sympa-thetic responses.

- **Electromyography**

  Electromyography is a technique used to measure the electrical activity produced by skeletal muscles. The generated signals can be used to determine the biomechanics of human movement. They are used to identify valence levels.

- **Electrocardiogram**

  ECG—The electrical activity of the heart is read using electrodes placed on the patient's body and the graph is recorded.

The data collected from these IoT devices is analyzed by a deep convolutional neural network (CNN) to determine the hidden emotional state of human emotions. In a treatise,[18] the author examined five basic emotions: happiness, sadness, relaxation, disgust, and neutrality. The paper argues that traditional emotion recognition uses facial recognition, speech recognition, cognitive-based mood analysis, and mood analysis from social networking sites, but these are inevitably disadvantageous. There is a point. In facial recognition, a person may be forging a facial expression, but the emotions covered are different. Language is not suitable for continuous sentiment analysis. Similarly, the person is not always connected to social networking sites. Therefore, the authors propose an innovative way to analyze emotions from the physiological state of the body by reading electrical signals generated by important parts of the body. This is because it forms the basis of all emotions that humans reflect. The IoMT sensors EDA, EMG, and ECG extract user arousal and value. The challenge of mapping these readings with human emotions is solved with the help of machine learning processes.

The framework referred to as IASM collects the combined data from all three sensors and predicts the affective state of mind. Collection of data from the sensors having varied parameters and converting the physiological reading to the psychological pattern is done with help of a series of the machine learning procedure. The signal peaks and index are identified through a state machine design. A number of features are extracted from the ECG, and the graph with peaks is recorded. Similarly, the skin conductance level is measured with several features and the graph of power spectral density against the happy and disgusted emotions is drawn. The joy or happiness and the valence level are measured by features extracted by the EMG like mean, sd, and integral (ZMA).

The ISAM framework has shown 90% accuracy in determining sad emotions. In general, it has achieved 87.5% classification accuracy compared to 72% of the radio-frequency-based emotion analyzer and 82.9% of SVM-based classifier.

7.5 CHALLENGES IN HEALTHCARE IOT

Every technology has its share of challenges. This section discusses the challenges of IoT in healthcare under three categories, technology-oriented challenges, the general people's outlook about remote healthcare, and the security issues of maintaining the patient's data and the real-time data as it involves the life of patients.

7.5.1 TECHNOLOGY-ORIENTED CHALLENGES

Internet disruption is the main issue faced by the IoT healthcare industry. The internet has to be seamlessly connected with maximum bandwidth; any interruptions may lead to adverse effects. The following are the technology-oriented challenges faced by Healthcare IoT.

7.5.1.1 RISK THE PATIENT'S LIFE

Internet disruptions can pay a very heavy price since the life of the patient is in line with the healthcare domain. Hence, zero tolerance toward the internet connectivity problem has to be ensured. An exhaustive test on the load, network bandwidth, and latency must be done for the worst-case scenarios in both mobile and web applications. And the peak load must be tested. The IoT crashes are unacceptable when life-saving sensitive equipment is involved. For instance, a Smart insulin pump used by patients releases insulin based on the current insulin level of the patient. It is considered a Digital Pancreas. Such equipment has no window to malfunction.[27]

7.5.1.2 INCORRECT RESULTS

Apparently, the stability of the user's internet connection may twist the test results. In case a patient's smart thermometer or inhaler faces a loss of connectivity or bandwidth issues, the receiver is likely to get an erroneous reading. To circumvent this, it is ensured that the readings are copied into local storage before being relayed to the destination.

7.5.1.3 NO PLANNED DOWNTIME

There is no place for planned downtime in IoT devices as it is seamlessly connected and relays data. So, the patches and updates need to be integrated alongside the device operation. Additionally, the patches should not tamper the sensitive data.

7.5.1.4 NEED FOR SPECIALIZED TOOL TO HANDLE DIVERSIFIED PROTOCOLS

As the IoT protocol lacks standardization, multiple IoT devices come up with multiple protocols. MQTT is the predominantly used protocol as it handles low-bandwidth and low-memory devices. However, few devices also employ HTTP, CoAP, XMPP, and many others. To enable communication between different devices with different protocols and with the servers, a fitting load test tool is brought into the picture by the performance test engineers. This is considered overhead.

7.5.1.5 REMOTE PLACES WITH A LACK OF INFRASTRUCTURE AND CONNECTIVITY

In developing and underdeveloped nations, the main challenge would be the infrastructure to get connected to the web.[2] In such cases, the remote healthcare service is a far-fetched concept.

7.5.2 ADAPTING TO REMOTE AND TELEHEALTH

People must adapt to embrace remote healthcare. Still, the major population holds inhibition toward remote healthcare and only trusts in personally visiting the hospital and doctors. However, the COVID-19 pandemic, on a positive note, has increased the number of people using remote healthcare. As the hospitals were closed for regular patients and illnesses, the people had to switch to telehealth and the records show that an overwhelming number of them are comfortable with it and will continue to use it.

7.5.3 DATA SECURITY

The data security in healthcare IoT is very much in the growing stage exposing the patients' life and privacy to undue risk.[28,30,31] The security vulnerabilities can be exploited by hackers with vested motives. The patient's primary concern is that his Health records should be confidential. The Patient Health records maintained centrally by the hospital can become a viable target for cyber-attacks. Many kinds of attacks, storage attacks, nodal attacks, and network attacks are possible in the IoT due to their complex design. An IoT framework involves lots of devices and interfaces, the sensor devices that communicate with themselves and with the storage node, with a gateway, and eventually to the cloud, thereby cyber-attacks are possible at various levels.

Medjacking is the term referring to medical device hijacking.[29] If a hacker gains access to a single device connected to the network, the hacker can control the network. The hacker also gets access to control the dosage of the drug to be given to a patient, which could also prove lethal.

The next section deals with the security threats and the different ways of combatting them.

7.6 SECURITY ISSUES AND DEFENSE MECHANISMS AND IOT

The crux of IoT is to bring all possible devices to relate to the internet seamlessly and make data transmission and real-time data analytics a standard practice. However, with it comes a great deal of security risk. In Healthcare, the accuracy and privacy of the data is pivotal. There always exist security threats in terms of invasion of privacy, unauthenticated access, denial of service, man-in-middle attacks, remote hijacking, and malware botnets. The three principles of security—Confidentiality, Integrity, and Authenticity must not be compromised.

7.6.1 SECURITY REQUIREMENTS IN HEALTHCARE IOT

7.6.1.1 CONFIDENTIALITY

It ensures the privacy of the information being passed from the sender to the receiver, additionally protecting the user's identity and protecting from external intrusion. The data passes through different nodes, storage nodes, processing nodes, IoT Gateway, and cloud. There are several interfaces and gateways through which the data must be transferred without compromising confidentiality. The risk is the disclosure of sensitive data or the sensitive data landing in neighboring nodes or unauthorized users.

7.6.1.2 INTEGRITY

Integrity is the method of ensuring that the data is real and accurate. The data received by the destination is as sent by the source. There should not be any unauthorized deletion, modification, or insertion of data. There are viable chances of data tapered by malignant users and significant errors due to channel imperfection or electromagnetic disturbances. It is inevitable that IoT uses secure mediums and interfaces.

7.6.1.3 AUTHENTICATION

Authentication is responsible to validate the identity of the source and the destination and that they are legitimate users. Unauthorized gain of access to data by malignant users is a major threat. Tag cloning, RFID eavesdropping, and spoofing are some of the common authentication attacks.

7.6.2 ATTACKS ON IOT DEVICES

Some of the common attacks that happen on IoT devices are discussed below. A plethora of security attacks are possible on IoT devices. A few are mentioned in the section.

7.6.2.1 SINKHOLE ATTACK

A sinkhole attack is a network layer attack. In this attack, a compromised or malicious node broadcast that it is the next best hop for all the nodes present in the network thereby rerouting all the data to it. Since it will never drop a packet, the presence of a sinkhole will remain unidentified. Such a node poses a huge threat to the data. Since it sinks all the data of the network it is called a Sinkhole attack.

7.6.2.2 BLACKHOLE ATTACK

In this attack, the compromised or the malicious node drops all the packets routed through it causing data loss and eventually stopping the data traffic. It is also a network attack.

7.6.2.3 SELECTING FORWARDING ATTACK (GRAYHOLE ATTACK)

This is a variant of a blackhole attack. In this attack, the compromised node does not drop all the packets. It selectively drops the packet and hence remains undetected by the Intrusion Detection System of the network.

7.6.2.4 WORMHOLE ATTACK

A wormhole attack is said to be a deadly attack. In this attack, two malicious nodes occupy a strategic position in the network and form a tunnel between them. The data received by one node is transferred through the tunnel to the other node. It can record all the data that passes through the network.

7.6.2.5 SYBIL ATTACK

It is a type of network attack where a single node carries multiple identities thereby causing chaos in the network. Sybil attacks result in contradicting routing paths and hence the data packets get lost, redundant, or disrupted. The effectiveness of fault tolerance is reduced.

7.6.2.6 DENIAL-OF-SERVICE ATTACK

The denial-of-service attack floods the network traffic with useless data thereby making the network unavailable to the authentic user traffic. This attack can happen both at the network and at application levels.

7.6.3 DEFENSIVE MECHANISM

A standard set of protocols is followed to protect the system from cyber-attack which includes IDS, firewall, encryption, antivirus, regular software updates, and access management. Currently, blockchain technology which is considered the state-of-the-art technology is also implemented in healthcare IoT. Fault tolerance is also one of the essential functions that are incorporated into the healthcare domain.

7.6.3.1 KEY MANAGEMENT

Key management plays an important role in cryptography. All data passing over the network must be encrypted. A cryptographic process (key generation, key sharing, encoding and decoding with public or private key cryptography) must be performed to prevent malicious activity during data transfer from source to destination.

7.6.3.2 USER/DEVICE AUTHENTICATION AND AUTHORIZATION

Authentication and authorization denote every user or device is given a unique id and login. When the device enters the network, the login of the device must be authenticated with the right credentials. Thereby any malicious attacker cannot easily hamper the network or any device connected to the network.

Authorization is about access rights and control. Each node and user is given only the required level of access to information. A server that monitors doctors' in and out time is not given access to patients' details. The access rights and privileges are predetermined and set, and it is associated with the login credentials.

7.6.3.3 INTRUSION DETECTION

An intrusion detection system (IDS) is installed on the network to monitor anomalous activity and send information to administrators to block malicious IP addresses.[30] Sinkhole, blackhole, and gray hole attacks all occur at compromised network nodes.

The functioning of an IDS is based on the following

- It identifies the sign of an intruder.
- It provides information about the location (i.e., suspected, IP address) of the intruder.
- It logs the information on ongoing activities.
- It tries to stop the malicious activities if they are detected.
- It then reports the information of malicious activities to the administrator (i.e., intrusion behavior that is either active attack or passive attack).
- It also provides information about types of intrusion (for example, which types of attack_mirai or echo bot).

7.6.3.4 FAULT TOLERANCE

- The main function of fault tolerance is to check data integrity. Fault-tolerant mechanisms retrieve data if it is lost in transit. In fog computing, storage nodes and processing nodes are different. As shown in Figure 7.4, there are multiple storage nodes connected to processing nodes, with various relay nodes in between. In such scenarios, some relay nodes may not work due to node failure, malfunction, or physical destruction of storage nodes. You should choose another path to connect to the processing node. Therefore, every storage node requires multiple possible paths to the processing node. In this way, the system does not have to wait for the failed node to recover, but instead redirects the packet through another node.[31]

When there is the possibility of rerouting, the data packages system should be equipped to trace back to the source. This is achieved by designating every storage node with a unique index and attaching a timestamp to the data packets. Every data packet will carry the index of the storage node along with a timestamp. Thereby, when a processing node receives a data packet, it can relate to the source and in case there has been rerouting with the help of the additional index and the timestamp it can trace back to the source. And if the data packets received are not in order, with the help of a timestamp the sequence of the data packets can be retrieved.

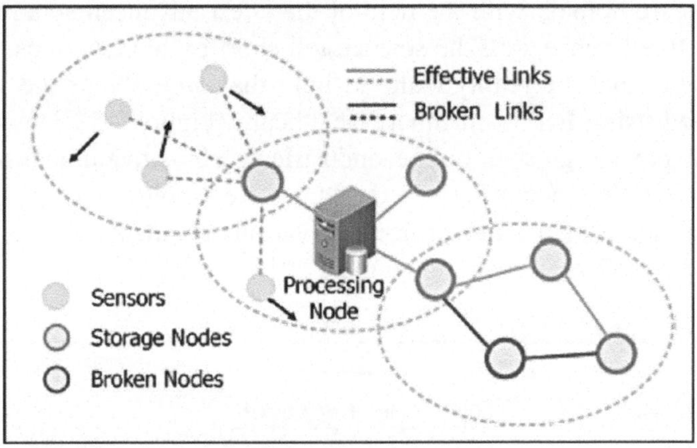

FIGURE 7.4 Implementation of fault tolerance in IoT nodes.

7.6.3.5 *BLOCKCHAIN TECHNOLOGY*

Blockchain technology can be called an electronic ledger. All transactions by users or devices connected to the blockchain are recorded and stored in the ledger, originally used to maintain cryptocurrencies—Bitcoin Ledger. The same technology is currently used in Healthcare IoT.[32] As mentioned in protecting the Internet of Things (IoMT) for Medical Things Using Private Blockchain Networks, a private blockchain network will be created within the Healthcare IoT network to provide another layer of verified authentication and security. Patient details are entered into the system along with medical history, eliminating the need for patients to carry paper copies of reports. Data is fully decentralized, avoiding outages and data loss and enabling faster recovery. All members of the group will be notified when a third party attempts to modify the database. Data is transparent, so changes made by one member of the group are propagated to everyone. You can provide an additional level of security for your data transmissions based on generally agreed security parameters. Database changes are accepted after approval by all members.

7.7 COVID 19—HOW IOT ROSE THE VALUE TO AVOID THE GLOBAL PANDEMIC

The world was swept by an unprecedented challenge in the form of a COVID-19 virus whose outbreak exposed the entire humankind to a pandemic situation.[33] Although widespread research has been happening in the healthcare domain with the help of the latest advancements in IoT and Artificial Intelligence, still the science fell short by huge bounds. It was in this situation that the reality came to light that only 5% of the work that is delivered using IoT is on Smart Healthcare (Figure 7.5). However, IoT did find explosive growth in other industries such as manufacturing, smart city, smart energy, and wearable consumer electronics. The figure shows the Global share of the IoT projects in various segments. The research is conducted by IoT Analytics showing the 2018 ranking.[34]

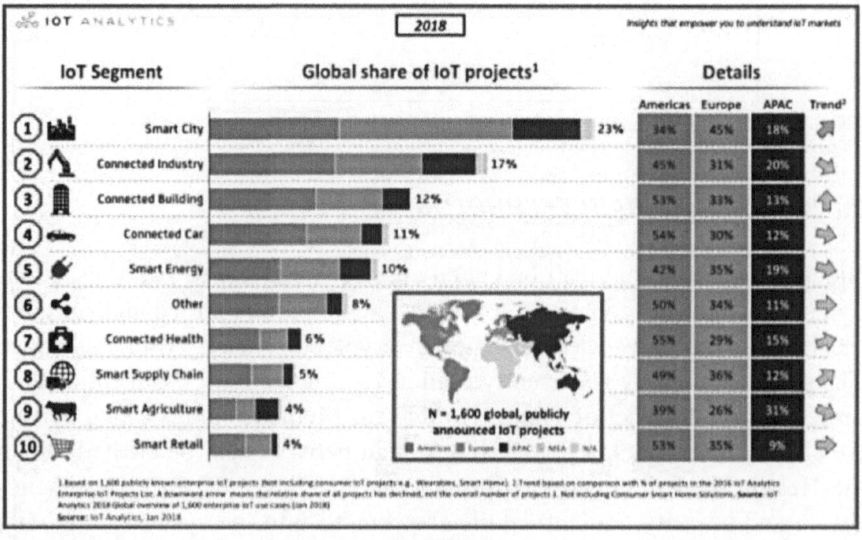

FIGURE 7.5 Top IoT segments SRC: IoT analytics.

Source: Reprinted from IoT Analytics. https://iot-analytics.com/top-10-iot-segments-2018-real-iot-projects/

About COVID-19

Coronavirus disease 2019 (COVID-19) is an acute respiratory disease caused by the new coronavirus (SARS-CoV-2), most often with droplet infections, direct contact with affected cases, and contamination infected by surface/object. The virus survives on the surface of the environment for various periods of time, but it is easily inactivated by chemical disinfectants.[33]

The deadly fact of COVID-19 is that patients can be infected, but they show no symptoms for the first 2 weeks, but can spread the disease. This asymptomatic spread is a challenge with social distancing emerging as the most viable solution. The spread of communities in a larger population and the fear of disease outbreaks put health facilities at risk as there is a large gap between supply and demand.

IoT helped to handle most of these challenges effectively.[35] Several government initiatives were carried out that involved IoT. A few of such initiatives are mentioned below.

7.7.1 DECODE THE OUTBREAK AND IDENTIFY PATIENT ZERO

The IoT with the help of the Geographic Information System (GIS) can help in leading the way to the patient zero or identifying and stopping community spread. The data from the infected patients' mobile when overlaid with the GIS will help in tracking movement and provide two-way information. On the upstream, patient zero can be detected by doing upstream research on how the patient caught the infection, where he traveled, whom all he met, etc., and on the downstream, it will help to identify who all comes in contact with the infected person.

7.7.2 QUARANTINED PATIENT CARE

Although IoT is very much into telemetry where the patient's crucial biometric data like heartbeat and blood pressure measurements are taken while the patient is at his residence or workplace and the data is sent to the medicos for constant monitoring, the COVID-19 posed new challenges. Anyone could be affected by the disease, the person without any history of medical conditions and who is not into any kind of monitoring can be affected by COVID-19.

With the crunch on the number of frontline healthcare workers and the risk of cross-infection with the patients, drones are used. A drone is made to fly to the quarantined patient's residence, all the patient needs to do is to come to his balcony, the drone will read the temperature with an infrared thermometer (Figure 7.6) and upload the data to the cloud.[34] The patient can update his further conditions using his mobile and the collective data can sum up the current condition of the patient.

FIGURE 7.6 Drone to monitor patients and deliver medicines.
Source: Reprinted from Netscribes. https://www.netscribes.com/drones-in-healthcare/

7.7.3 PUBLIC SURVEILLANCE

Drones are employed for public surveillance to ensure quarantine, the effectiveness of curfew, and social distancing across the globe.[34] Gathering in groups is a potential threat; the drones enabled with IoT sensors are used to check on such gatherings. Screening people whether they are wearing masks, exhibiting abnormal conditions, and temperature recording of people in a crowd are practiced.

7.7.4 SAFEGUARDING HYGIENE

A new startup, SoapyCare,[36] has developed a so-called smart sink. With the slogan "Your hygiene is in your hands," these stand-alone microstations (similar to water dispensers) use IoT sensors to make soaps or disinfectants that need to be used. Dispense the exact amount of water, as well as the amount and temperature of the water, which requires hand washing at the right time. With a variety of settings recommended for homes, gastronomy, day care centers, or senior centers, these hand-washing stations ensure hygiene and ultimately your hygiene. Soapy Care systems have already been seen in the United States, Africa, and Asia, and their phones have not stopped ringing since the outbreak of COVID-19.

7.7.5 IOT AND ROBOTICS

To reduce the efforts of overstrained frontline healthcare workers, the contagion robots (Figure 7.8) are used in delivering multiple services at the hospitals.

- To deliver food and medicines to the patients.
- Monitoring patients' body temperatures.
- For disinfecting the designated areas.
- To provide video conferencing between doctors and patients.
- To detect people without masks.

7.7.6 THE SMART DISINFECTION AND SANITATION TUNNEL

Disinfectant tunnels were placed in public places where a person needs to walk through the tunnel which is enabled with an IoT sensor-based disinfectant spray. The right proportion of the disinfectant chemical is sprayed at the required amount as the person passes through the tunnel, thereby killing the virus if the person has been contacted anywhere by chance (Figure 7.9).

FIGURE 7.7 Disinfectant tunnels.

Source: Authors

The drones were also used to spray disinfectants to places that cannot be reached by people or where healthcare workers are in shortage.

Going forward, IoT in the healthcare domain will see a mammoth boost after the COVID-19 crisis has exposed inefficient processes and technology bottlenecks. The organizations that came up with ad-hoc fixes would want to streamline their inventions in the future.

7.7.7 SMART MASKS AND SMART MEDICAL EQUIPMENT

Wearing a mask will prevent the spread of infection and prevent individuals from contracting airborne infectious droplets. The masks are designed with some intelligence which makes them smart masks.[37] They are built with features to inactivate the span of virus-laden aerosols in a bidirectional way. It has sensors to assess the health of the patient and some masks are also designed to amplify the voice.

7.8 THE FUTURE OF THE IOT IN HEALTHCARE

IoT is one of the keystone players that drive the future of healthcare. The future of IoT depends very much on its enabling technologies, one of which is wireless communication. The latest milestone achieved in wireless communication is 5G. It is revered as the flagship of wireless long-distance communication. The combination of IoT and 5G is expected to pave the way for explosive growth in the healthcare domain. Another enabling technology is data analytics. Artificial intelligence with its deep learning technology has a promising outlook for the IoT in healthcare.[38]

7.8.1 IOT AND 5G

With the rollout of Fifth Generation (5G) round the corner, many innovations, and blueprints to match IoT for the 5G revolution are in pipeline. 5G is expected to undermine the full potential of IoT. 5G will spur innovations across many industries making IoT an integral part of our economy and lifestyle. It is expected that there would be 550 million 5G subscriptions within 2 years of the rollout.

5G is said to be 100 times faster than 4G. A key differentiator is low latency combined with fast connectivity and ubiquitous coverage to accelerate 5G. The current challenges in IoT healthcare environments can be easily addressed with 5G. At the same time, a new definition of "telemedicine" without delays will emerge. With 5G, patient monitoring and critical clinical procedures in the fraction of a second will be flawless.

The 5G is capable of streaming a 4k video without buffering. So any application that requires real-time monitoring can get rid of local storage. The high-definition videos can be live telecasted without a fraction of a second delay in transmission. In other words, zero latency. We may imagine a surgery happening smoothly with the medical team and the patient thousands of miles apart.

7.8.2 IOT AND ARTIFICIAL INTELLIGENCE

Both Artificial Intelligence (AI) and IoT are booming technologies with inventions happening at a rapid rate. The IoT gets a new facelift with every new invention in the AI and vice versa.[38] As the analogy holds Internet of Things is the nervous system and the AI is the brain that controls and acts upon the nervous system; the combination of these two powerful paradigms provides limitless innovations.

Machine learning and deep learning the two powerful pillars of AI help IoT in a variety of areas. Business intelligence is applied to the huge cloud of data generated and stored by IoT devices. The AI identifies patterns, detects anomalies in the data that is generated by sensors, and makes predictions with high accuracy. As described in Ref. [39], through cognitive smart healthcare for pathology detection and monitoring, the patient EEG pathology is monitored and sent to the cloud where the deep learning is applied, and real-time decisions are made by the classification of the data whether they are normal or require emergency help.

Other AI technologies, such as speech recognition and computer vision, can help extract insights from data that previously required human verification. AI applications for IoT can help businesses avoid unplanned downtime, increase operational efficiency, develop new products and services, and improve risk management.

7.9 CONCLUSION

The IoT in healthcare is still considered to be in the emerging stage, and countless developments are expected in the near future. With the advent of 5G and further research and development in IoT-enabled technologies such as wireless communications, cloud storage, and big data analytics, IoT is expected to grow explosively. Healthcare IoT is in increasing demand due to the current pandemic situation of redefining healthcare with remote diagnostics, monitoring, and connected devices. This chapter described different application and implementation models, new fog and edge computing beyond traditional cloud models, the main challenges facing IoT in healthcare, and security practices and issues in IoT platforms. This chapter also describes common security attacks that occur in IoT environments. In addition, this chapter reveals how the IoT was launched to address the current pandemic situation in the fight against the deadly COVID-19 virus. Finally, the future growth lines of IoT, including realization technologies, were discussed.

KEYWORDS

- **IoT**
- **healthcare**
- **ecosystem**
- **automated manufacturing**
- **security**

REFERENCES

1. Habibzadeh, H.; Dinesh, K.; Rajabi Shishvan, O.; Boggio-Dandry, A.; Sharma, G.; Soyata, T. A Survey of Healthcare Internet of Things (HIoT): A Clinical Perspective. *IEEE Int. Things J.* **2020,** *7* (1), 53–71. DOI: 10.1109/JIOT.2019.2946359.
2. Poyner, I. K.; Sherratt, R. S. In *Improving Access to Healthcare in Rural Communities — IoT as Part of the Solution*, 3rd IET International Conference on Technologies for Active and Assisted Living (TechAAL 2019); London, UK, 2019; pp 1–6. DOI: 10.1049/cp.2019.0104.
3. Pathinarupothi, R. K.; Durga, P.; Rangan, E. S. IoT-Based Smart Edge for Global Health: Remote Monitoring With Severity Detection and Alerts Transmission. *IEEE Int. Things J.* **2019,** *6* (2), 2449–2462, April 2019.
4. https://www.who.int/gho/health_workforce/physicians_density/en/
5. Sarkar, S.; Saha, R. In *A Futuristic IOT Based Approach for Providing Healthcare Support Through E-Diagnostic System in India*, 2017 Second International Conference on Electrical, Computer and Communication Technologies (ICECCT), Coimbatore, 2017; pp 1–7. DOI: 10.1109/ICECCT.2017.8117810.

6. Ganesan, M.; Sivakumar, N. In *IoT Based Heart Disease Prediction and Diagnosis Model for Healthcare Using Machine Learning Models*, 2019 IEEE International Conference on System, Computation, Automation and Networking (ICSCAN), Pondicherry, India, 2019; pp 1–5. DOI: 10.1109/ICSCAN.2019.8878850.
7. Bisio, I.; Garibotto, C.; Lavagetto, F.; Sciarrone, A. When eHealth Meets IoT: A Smart Wireless System for Post-Stroke Home Rehabilitation. *IEEE Wireless Commun.* **2019,** *26* (6), 24–29. DOI: 10.1109/MWC.001.1900125.
8. Nooruddin, S.; Islam, M. M.; Sharna, F. A. An IoT Based Device-Type Invariant Fall Detection System. *IoT* **2020,** *9*, 100103.
9. Al-Mahmud, O.; Khan, K.; Roy, R.; Mashuque Alamgir, F. In *Internet of Things (IoT) Based Smart Health Care Medical Box for Elderly People*, 2020 International Conference for Emerging Technology (INCET), Belgaum, India, 2020; pp 1–6. DOI: 10.1109/INCET49848.2020.9153994.
10. Tsai, H.; Tseng, C. H.; Wang, L.; Juang, F. In *Bidirectional Smart Pill Box Monitored Through Internet and Receiving Reminding Message from Remote Relatives*, 2017 IEEE International Conference on Consumer Electronics - Taiwan (ICCE-TW), Taipei, 2017; pp 393–394. DOI: 10.1109/ICCE-China.2017.7991161.
11. Serdaroglu, K.;Uslu, G.; Baydere, S. In *Medication Intake Adherence with Real Time Activity Recognition on IoT*, 2015 IEEE 11th International Conference on Wireless and Mobile Computing, Networking and Communications (WiMob), Abu Dhabi, 2015; pp 230–237. DOI: 10.1109/WiMOB.2015.7347966.
12. Autobed : https://sites.gatech.edu/hrl/autobed-a-web-controlled-robotic-bed/
13. Garg, U.; Ghanshala, K. K.; Joshi, R. C.; Chauhan, R. In *Design and Implementation of Smart Wheelchair for Quadriplegia patients using IOT*, 2018 First International Conference on Secure Cyber Computing and Communication (ICSCCC), Jalandhar, India, 2018; pp 106–110. DOI: 10.1109/ICSCCC.2018.8703354.
14. Rajith, A.; Soki, S.; Hiroshi, M. In *Real-Time Optimized HVAC Control System on Top of an IoT Framework*, 2018 Third International Conference on Fog and Mobile Edge Computing (FMEC), Barcelona, 2018; pp 181–186. DOI: 10.1109/FMEC.2018.8364062.
15. Amin, S. U.; Hossain, M. S.; Muhammad, G.; Alhussein, M.; Rahman, M. A. Cognitive Smart Healthcare for Pathology Detection and Monitoring. *IEEE Access* **2019,** *7*, 10745–10753.
16. Thierry, E. Internet of Things in Emergency Medical Care and Services, 2018. 10.5772/intechopen.76974.
17. Woo, M. W.; Lee, J. W.; Park, K. H. A Reliable IoT System for Personal Healthcare Devices. *Future Gener. Comput. Syst.* **2018,** *78* (2), 626–640.
18. Alam, M. G. R.; Abedin, S. F.; Moon, S. I.; Talukder, A.; Hong, C. S. Healthcare IoT-Based Affective State Mining Using a Deep Convolutional Neural Network. *IEEE Access* **2019,** *7*, 75189–75202.
19. Leng, J.; Lin, Z.; Wang, P. In *Poster Abstract: An Implementation of an Internet of Things System for Smart Hospitals*, 2020 IEEE/ACM Fifth International Conference on Internet-of-Things Design and Implementation (IoTDI), Sydney, Australia, 2020; pp 254–255. DOI: 10.1109/IoTDI49375.2020.00034.
20. Ahmadi, V.; Benjelloun, S.; El Kik, M.; Sharma, T.; Chi, H.; Zhou, W. In *Drug Governance: IoT-based Blockchain Implementation in the Pharmaceutical Supply Chain*, 2020 Sixth International Conference on Mobile And Secure Services (MobiSecServ), Miami Beach, FL, USA, 2020; pp 1–8. DOI: 10.1109/MobiSecServ48690.2020.9042950.

21. Saravanan, M.; Marks, A. M. In *MEDIBOX — IoT Enabled Patient Assisting Device*, 2018 IEEE 4th World Forum on Internet of Things (WF-IoT), Singapore, 2018; pp 213–218. DOI: 10.1109/WF-IoT.2018.8355207.
22. Sood, S. K.; Mahajan, I. IoT-Fog-Based Healthcare Framework to Identify and Control Hypertension Attack. *IEEE IoT J.* **2019,** *6* (2), 1920–1927.
23. Wang, K.; Shao, Y.; Xie, L.; Wu, J.; Guo, S. Adaptive and Fault-Tolerant Data Processing in Healthcare IoT Based on Fog Computing. *IEEE Trans. Netw. Sci. Eng.* **2020,** *7* (1), 263–273.
24. Pace, P.; Aloi, G.; Gravina, R.; Caliciuri, G.; Fortino, G.; Liotta, A. An Edge-Based Architecture to Support Efficient Applications for Healthcare Industry 4.0. *IEEE Trans. Ind. Inf.* **2019,** *15* (1), 481–489.
25. Subahi, A. F. Edge-Based IoT Medical Record System: Requirements, Recommendations and Conceptual Design. *IEEE Access* **2019,** *7*, 94150–94159.
26. Khan, M. A. An IoT Framework for Heart Disease Prediction Based on MDCNN Classifier. *IEEE Access* **2020,** *8*, 34717–34727.
27. Sargunam, B.; Anusha, S. In *IoT Based Mobile Medical Application for Smart Insulin Regulation*, 2019 IEEE International Conference on Electrical, Computer and Communication Technologies (ICECCT), Coimbatore, India, 2019; pp 1–5. DOI: 10.1109/ICECCT.2019.8869227.
28. Ahanger, T. A.; Aljumah, A. Internet of Things: A Comprehensive Study of Security Issues and Defense Mechanisms. *IEEE Access* **2019,** *7*, 11020–11028, 2019.
29. https://en.wikipedia.org/wiki/Medical_device_hijack
30. Wazid, M.; Das, A. K.; Rodrigues, J. J. P. C.; Shetty, S.; Park, Y. IoMT Malware Detection Approaches: Analysis and Research Challenges. *IEEE Access* **2019,** *7*, 182459–182476. DOI: 10.1109/ACCESS.2019.2960412.
31. Albahri, O. S., et al. Fault-Tolerant mHealth Framework in the Context of IoT-Based Real-Time Wearable Health Data Sensors. *IEEE Access* **2019,** *7*, 50052–50080 DOI: 10.1109/ACCESS.2019.2910411.
32. AnithaKumari, K.; Padmashani, R.; Varsha, R.; Upadhayay, V. Securing Internet of Medical Things (IoMT) Using Private Blockchain Network. In *Principles of Internet of Things (IoT) Ecosystem: Insight Paradigm*; Intelligent Systems Reference Library; Peng, S. L., Pal, S., Huang, L., Eds.; Springer, 2020; vol 174.
33. https://www.who.int/emergencies/diseases/novel-coronavirus-2019
34. https://iot-analytics.com/top-10-iot-segments-2018-real-iot-projects/
35. Chamola, V.; Hassija, V.; Gupta, V.; Guizani, M. A Comprehensive Review of the COVID-19 Pandemic and the Role of IoT, Drones, AI, Blockchain, and 5G in Managing its Impact. *IEEE Access* **2020,** *8*, 90225–90265. DOI: 10.1109/ACCESS.2020.2992341.
36. Smart Sink: https://www.calcalistech.com/ctech/articles/0,7340,L-3799860,00.html
37. https://www.theengineer.co.uk/smart-ppe-face-masks-sensors-northwestern/
38. Mohanta, B.; Das, P.; Patnaik, S. In *Healthcare 5.0: A Paradigm Shift in Digital Healthcare System Using Artificial Intelligence, IOT and 5G Communication*, 2019 International Conference on Applied Machine Learning (ICAML), Bhubaneswar, India, 2019; pp 191–196. DOI: 10.1109/ICAML48257.2019.00044.
39. Amin, S. U.; Hossain, M. S.; Muhammad, G.; Alhussein, M.; Rahman, M. A. Cognitive Smart Healthcare for Pathology Detection and Monitoring. *IEEE Access* **2019,** *7*, 10745–10753.

CHAPTER 8

Transmission Dynamics Prediction Models with Data Visualization and Analysis of COVID-19

PRANATI RAKSHIT, PRONIT SARKAR, and SHANKAR DEBNATH

Computer Science and Engineering Department,
JIS College of Engineering, Kalyani, India

ABSTRACT

SARS-CoV-2 (n-coronavirus), as we already know, is a new virus in this family, first introduced in 2019 and changed the lives of everyone around the world. This current work focuses on data visualization and analysis using predictive models. The reason is that the data itself speaks volumes when properly visualized and analyzed. Management of this disease varies from country to country. I'm analyzing data from COVID-19, getting confirmation, recovery, and death data for three different time periods, and would like to gain insight into the rules and conditions in different countries during different time periods. Another part of this current work is the prediction or prediction of confirmed, recovered, and fatalities. We got good R-squared values close to 1 for our predictive model, which is a satisfying result. We used Linear Regressor, K-Nearest Neighbor Regressor, and Gradient Boosting Regressor as predictive models and compared their performance. Best results were obtained with the K-Nearest Neighbor Regressor.

8.1 INTRODUCTION

Now everyone is familiar with the name coronavirus or COVID-19.[1] It made history after being introduced to Wuhan city.[2] Now it is gradually reaching almost the whole world. All countries are fighting this disease and trying

Global Digital Transformation and the Covid-19 Pandemic.
Rajdeep Chowdhury, K. Martin Sagayam, Neyara Radwan, Ahmad A. Elngar (Eds.)

to get rid of this unprecedented disease. Some countries do not have such advanced medical equipment, lack of medical personnel due to the limitation of the medical system, and have many problems such as large-scale outbreaks of diseases.[3] Coronaviruses are basically a family of viruses that can cause underlying illness and vary from person to person with a variety of symptoms, from mild to severe. Such severe cases were Middle East Respiratory Syndrome (MERS-CoV) and Severe Acute Respiratory Syndrome (SARS-CoV).[4–6] SARS-CoV-2 (n-coronavirus), as we already know, is a new virus in this family, first introduced in 2019,[7] and people throughout the universe literally have problems. Colds and coughs are common symptoms and can be complicated and transformed into cytokine storms, which have never been observed in humans before.[8] The first case of the virus was detected in November 2019 in Wuhan, China, a city with a population of 1.1 million.[9] After some time, it was declared a pandemic by WHO due to its high rate of worldwide spread.[10] Pandemic is spreading all over the world. Understanding this spread is therefore of great interest. Also, it is very important to know the dynamics of the spread to help find some reasons for high and relatively low-impact areas. In this chapter, we use three periods to analyze the propagation dynamics. This work is an attempt to analyze cumulative data of confirmed, fatal, and recovered cases over time. This current work focuses on analyzing and analyzing the prevalence of this virus around the world.

This study aims to establish a basis for a causal relationship between pandemic severity in the country and government control. Establishing such relationships can provide governments with important insights that can be used to enact effective policies and laws to combat pandemics. The data used in the analysis performed is based on publicly available statistics and government-reported statistics on her COVID-19 patients in each country. The main dataset used is from the Johns Hopkins University Center for Systems Science and Engineering (CSSE). This dataset consisted of data collected from a variety of sources, including WHO, the European Centers for Disease Control and Prevention, the US CDC, and various other government and nongovernmental organizations. In Section 8.2 of this chapter, we discuss the literature search with motivation. Section 8.3 describes the data visualization and analysis, Section 8.4 describes the predictive model, Section 8.5 describes the results and discussion, and Section 8.6 summarizes the study.

8.2 LITERATURE SURVEY

This section describes some literature reviews and attempts to find the limits or scope of the research in this chapter. Yadav[11] used regression analysis based on six models for the purpose of data analysis of the COVID-19 dataset. They mainly worked on Indian datasets. Their 6th degree polynomial regression model is designed to help physicians and governments consider and formulate future plans and precautions. Dey et al.[12] compiled and analyzed occurrence information in seven open datasets. They performed an exploratory data analysis and made visualizations based on the number of reported confirmed cases, deaths, and recovered cases. Mishra et al.[13] worked on machine learning to develop predictive models for confirmed, fatal, and recovered cases. A prediction error has occurred. The RMSE is 11225.84 and the forecast MAE is 8550.19. Khan et al.,[14] discussed the worldwide spread of COVID-19 as one of the most dangerous diseases. They used CoroNet, a deep CNN method based on the Xception architecture. It can detect COVID-19 in its early stages from a patient's chest X-ray. Several studies have been conducted on COVID19 detection using machine learning and deep learning.[14–23] And a lot of research has been done on transfer learning.[24–30] Motivation: Some work has been done using machine learning and transfer learning to detect COVID-19. However, little work has been done on data analysis and visualization. As such, this work focuses primarily on data visualization and its analysis to gain insight into the data. Another part is the predictive model. This model has received little work in the existing literature. The goal is to get better prediction results.

8.3 DATA VISUALIZATION AND ANALYSIS

Here, for data analysis, we have taken data from John Hopkins. The current status can be seen as we have taken data up to April 24, 2021. Since the dataset is updated automatically by the CSSE, here the live situation of COVID-19 can be seen at the global level. We have analyzed the data in three-time spans and compared or analyzed accordingly.

TABLE 8.1 Confirmed Cases.

Date	Number
August 17, 2020	21,940,380
December 19, 2020	75,819,239
April 24, 2021	145,843,876

TABLE 8.2 Death Cases.

Date	Number
August 17, 2020	775,348
December 19, 2020	1,677,957
April 24, 2021	3,090,184

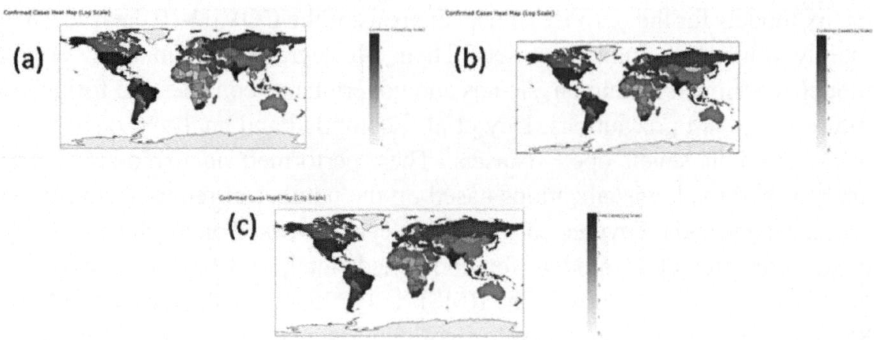

FIGURE 8.1 Confirmed cases as on (a) August 17, 2020; (b) August 19, 2020; and (c) April 24, 2021.

TABLE 8.3 Recovery Cases.

Date	Number
August 17, 2020	13,924,141
December 19, 2020	42,800,737
April 24, 2021	84,002,731

Figure 8.1 explains the heat map of Confirmed cases in the world Map on (a) August 17, 2020; (b) December 19, 2020; and (c) April 24, 2021, respectively. Figure 8.2 explains the heat map of Death cases in the world Map on (a) August 17, 2020; (b) August 19, 2020; and (c) April 24, 2021, respectively.

Figure 8.3 explains the heat map of Recovery cases in the world Map on August 17, 2020; December 19, 2020; and April 24, 2021, respectively. Figures 8.4–8.6 are depicting the top ten countries on the basis of Confirmed cases on August 17, 2020; December 19, 2020; and April 24, 2021, respectively. From these figures, we can see that India came in 2nd position from 3rd position as per confirmed cases on December 19, 2020. The countries like South Africa, Peru, Mexico, Colombia, and Chile have improved their situation and they are excluded from the top ten confirmed cases on December 19, 2021. But France, Turkey, Italy, and Argentina came in the top ten on December 19, 2020.

S, India, and Brazil are in the same position on April 24, 2021 as like on December 19, 2021. Germany has come up in the top ten on the third date (April 24, 2021). Figures 8.7–8.9 are depicting the top ten countries on the basis of death cases on

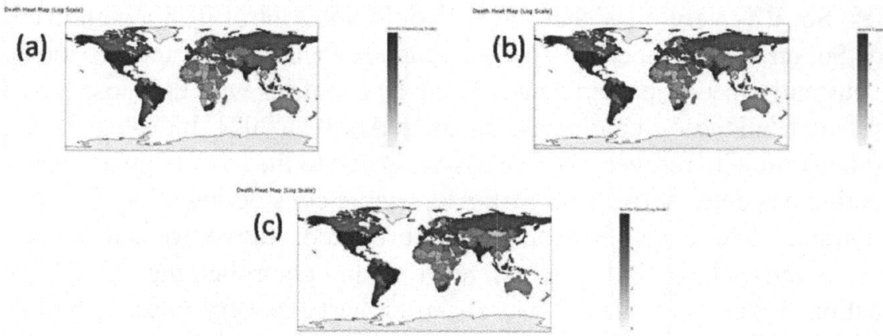

FIGURE 8.2 Death cases as on (a) August 17, 2020; (b) August 19, 2020; and (c) April 24, 2021.

August 17, 2020 and December 19, 2020 and April 24, 2021, respectively. From these first two figures, India came in 3rd position from 4th position as per death cases as on the 2nd date (December 19, 2021). But again India has improved its situation and restored its 4th position as on April 24, 2021. The countries like Peru have improved their situation and they are excluded from the top ten Death cases as on December 19, 2021 and also on April 24, 2021. France, Italy, and Iran are in the top ten on August 17, 2020 as well as December 19, 2020 but Iran has improved to be out of the top ten list on April 24, 2021. But Russia was not in the top ten on August 17, 2020 but came to the top ten on December 19, 2020 and April 24, 2021, respectively. Figures 8.10–8.12 are depicting the top ten countries on the basis of death cases on August 17, 2020, December 19, 2020, and April 24, 2021, respectively. From these first two figures, we can analyze that the US has adopted some better policies and it came to 1st position in recovery rate on December 19, 2020 which was previously in 3rd position on August 17, 2020. The countries like South Africa, Peru, and Chile are not in the top ten on December 19, 2020. It does not mean that they are in a bad position. But actually, they have improved their situation and they are excluded from the top ten confirmed cases also. That's why they are not top ten Recovery cases on December 19, 2020. India is in the topmost position with respect to Recovery cases as on April 24, 2021. Figures 8.13–8.15 are depicting the top ten countries on the basis of Active cases on August 17, 2020 and December 19, 2020, and April 24, 2021, respectively. From these first two figures, we can analyze that India, Peru, and Bangladesh are in a good situation according to active cases on December 19, 2020 as compared to the situation on August 17, 2020. The US is in the first position on both dates. So, it is very clear that the US could not control it in a significant way. So, up to December 19, 2020, we can see that the most active patients are currently in Europe and America, all 10 countries with the most active cases are from these 2 continents. But as on April 24, 2021, India has got the first position with respect to active cases. So due to the large population, the situation has deteriorated very fast. Also, while considering all 4 comparative graphs, that is, Confirmed, Death, Recovered, Active, we can see that some countries have done impressive work, and controlled the COVID-19 situation. India has achieved one of the highest recovery rates. India has recorded 13M+ cases which have recovered so far, which is remarkable, considering the population and social economic situation in which most of our population leaves. From the above, we can conclude that despite all the steps taken by the concerned govt, the situation has only worsened. The number of confirmed cases and deaths have increased many folds in an 8-month span.

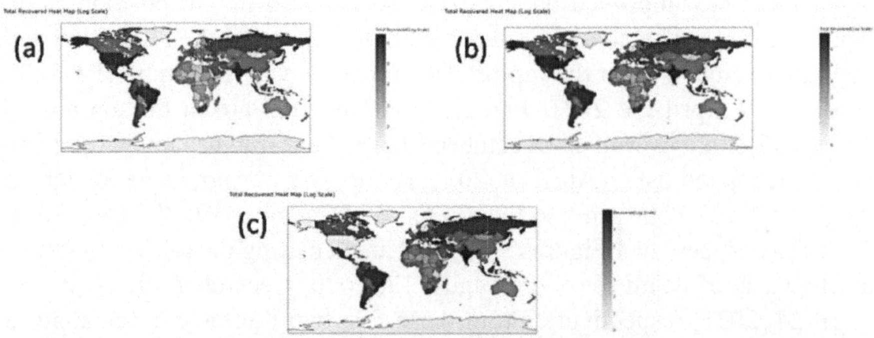

FIGURE 8.3 Recovery cases as on (a) August 17, 2020; (b) August 19, 2020; and (c) April 24, 2021.

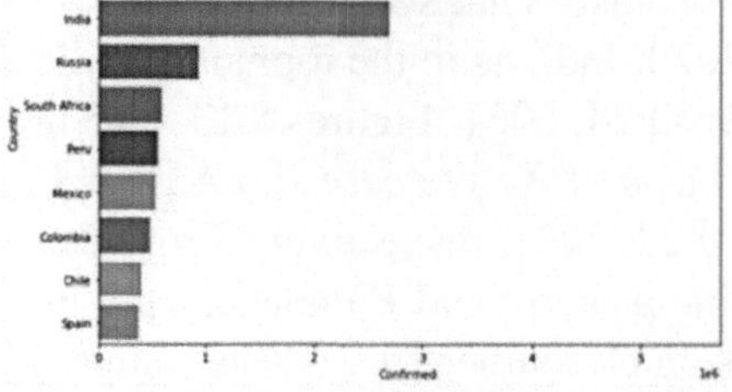

	Country	Confirmed	Deaths	Recovered	Active	Incident_Rate
174	US	5452115	170824	1865580	3415723	1654.832461
23	Brazil	3359570	108536	2699080	551954	1580.532245
79	India	2702742	51797	1977779	673166	195.850247
140	Russia	930276	15836	741045	173395	637.461502
156	South Africa	589886	11982	477671	100233	994.602983
134	Peru	541493	26481	370717	144295	1642.289000
113	Mexico	525733	57023	430840	37870	411.396506
37	Colombia	476660	15372	301525	159763	936.778662
35	Chile	387502	10513	360385	16604	2027.086019
158	Spain	364196	28670	150376	185150	778.949183

FIGURE 8.4 Top ten countries with respect to confirmed cases as on August 17, 2020.

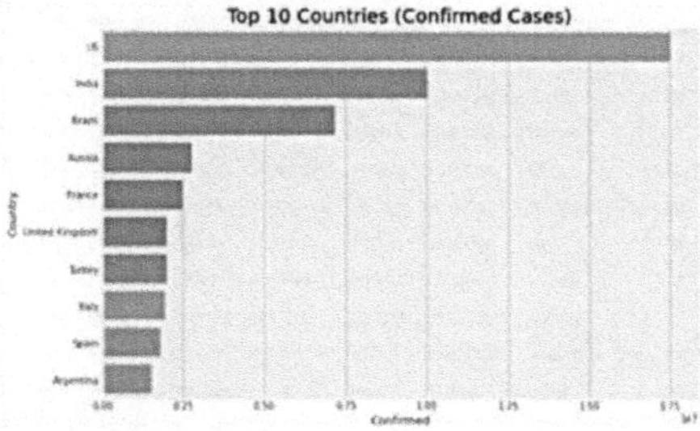

	Country	Confirmed	Deaths	Recovered	Active	Incident_Rate
177	US	17472522	313764	10397829	7176107	5303.280761
79	India	10004599	145136	9550712	308751	724.968638
23	Brazil	7162978	185650	6322955	654373	3369.871056
141	Russia	2792615	49744	2233060	509811	1913.609027
62	France	2499739	60347	188488	2250559	3829.637664
181	United Kingdom	1983017	66643	4340	1912034	2921.098434
176	Turkey	1982090	17610	1753552	210928	2350.144566
85	Italy	1921778	67894	1226086	627798	3178.498010
161	Spain	1797236	48926	150376	1597934	3843.961804
6	Argentina	1531374	41672	1356755	132947	3388.312143

FIGURE 8.5 Top ten countries with respect to confirmed cases as on December 19, 2020.

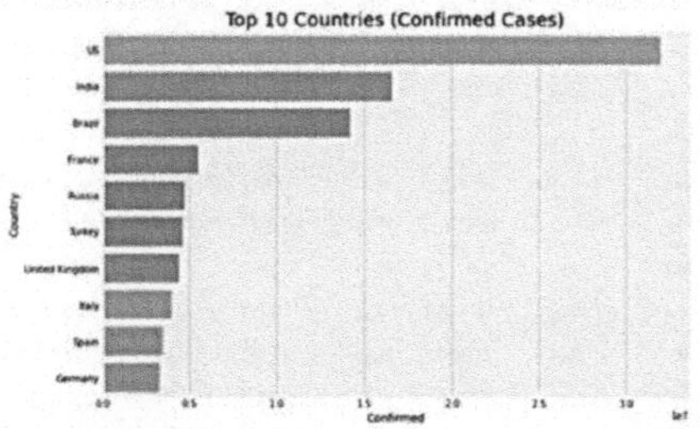

	Country	Confirmed	Deaths	Recovered	Active	Incident_Rate
178	US	32000794	571285	0	0	9712.919243
79	India	16610481	189544	13867997	2552940	1203.654219
23	Brazil	14237078	386416	12593957	1256705	6697.928860
62	France	5502171	102656	336234	5063134	8429.408548
142	Russia	4699988	106108	4330610	263270	3220.615611
177	Turkey	4591416	38011	10397829	7176107	5443.996671
182	United Kingdom	4418709	127670	14388	4276651	6509.013257
85	Italy	3949517	119021	3369048	461448	6532.248744
162	Spain	3468617	77591	150376	3240650	7418.742592
66	Germany	3276871	81542	2892350	302979	3911.096571

FIGURE 8.6 Top ten countries with respect to confirmed cases as on April 24, 2021.

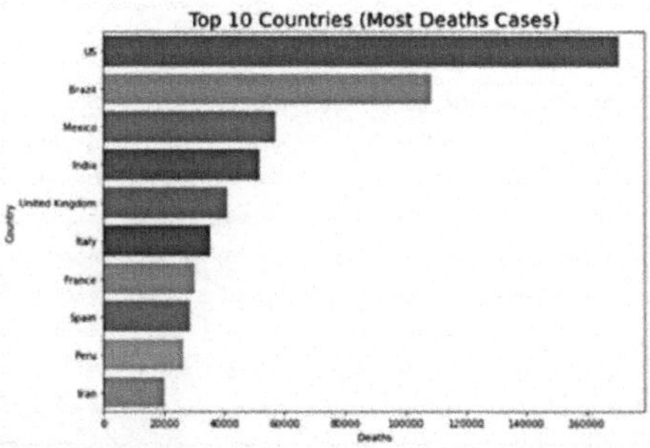

	Country	Confirmed	Deaths	Recovered	Active	Incident_Rate
174	US	5452115	170824	1865580	3415723	1654.832461
23	Brazil	3359570	108536	2699080	551954	1580.532245
113	Mexico	525733	57023	430840	37870	411.396506
79	India	2702742	51797	1977779	673166	195.850247
178	United Kingdom	322160	41466	1492	279202	474.560264
85	Italy	254636	35405	204142	15089	421.151673
62	France	256534	30434	84214	141886	393.013938
158	Spain	364196	28670	150376	185150	778.949183
134	Peru	541493	26481	370717	144295	1642.289000
81	Iran	347835	19972	300881	26982	414.124028

FIGURE 8.7 Top ten countries with respect to death cases as on August 17, 2020.

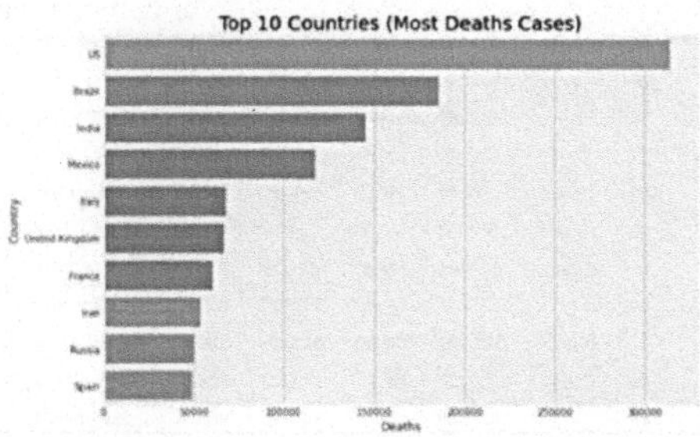

	Country	Confirmed	Deaths	Recovered	Active	Incident_Rate
177	US	17472522	313764	10397829	7176107	5303.280761
23	Brazil	7162978	185650	6322955	654373	3369.871056
79	India	10004599	145136	9550712	308751	724.968638
114	Mexico	1301546	117249	962397	221900	1018.485576
85	Italy	1921778	67894	1226086	627798	3178.498010
181	United Kingdom	1983017	66643	4340	1912034	2921.098434
62	France	2499739	60347	188488	2250559	3829.637664
81	Iran	1152072	53448	875943	222681	1371.629356
141	Russia	2792615	49744	2233060	509811	1913.609027
161	Spain	1797236	48926	150376	1597934	3843.961804

FIGURE 8.8 Top ten countries with respect to death cases as on December 19, 2020.

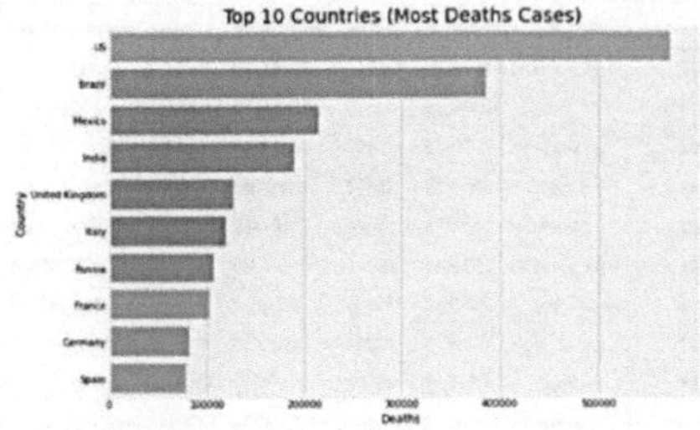

	Country	Confirmed	Deaths	Recovered	Active	Incident_Rate
178	US	32000794	571285	0	0	9712.919243
23	Brazil	14237078	386416	12593957	1256705	6697.928860
114	Mexico	2323430	214504	1845008	263918	1818.130087
79	India	16610481	189544	13867997	2552940	1203.654219
182	United Kingdom	4418709	127670	14388	4276651	6509.013257
85	Italy	3949517	119021	3369048	461448	6532.248744
142	Russia	4699988	106108	4330610	263270	3220.615611
62	France	5502171	102656	336234	5063134	8429.408548
66	Germany	3276871	81542	2892350	302979	3911.096571
162	Spain	3468617	77591	150376	3240650	7418.742592

FIGURE 8.9 Top ten countries with respect to death cases as on April 24, 2021.

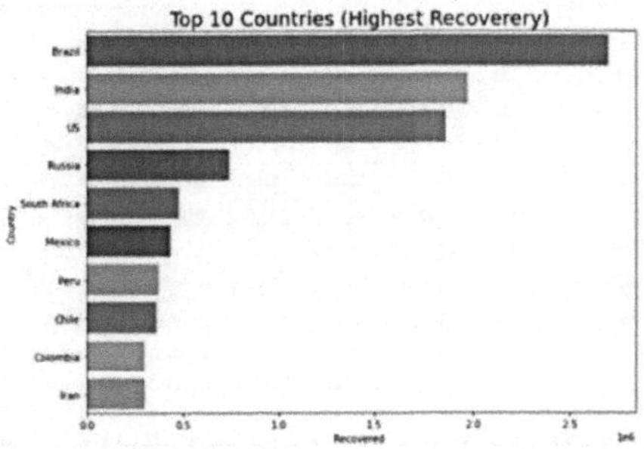

	Country	Confirmed	Deaths	Recovered	Active	Incident_Rate
23	Brazil	3359570	108536	2699080	551954	1580.532245
79	India	2702742	51797	1977779	673166	195.850247
174	US	5452115	170824	1865580	3415723	1654.832461
140	Russia	930276	15836	741045	173395	637.461502
156	South Africa	589886	11982	477671	100233	994.602983
113	Mexico	525733	57023	430840	37870	411.396506
134	Peru	541493	26481	370717	144295	1642.289000
35	Chile	387502	10513	360385	16604	2027.086019
37	Colombia	476660	15372	301525	159763	936.778662
81	Iran	347835	19972	300881	26982	414.124028

FIGURE 8.10 Top ten countries with respect to Recovery cases as on August 17, 2020.

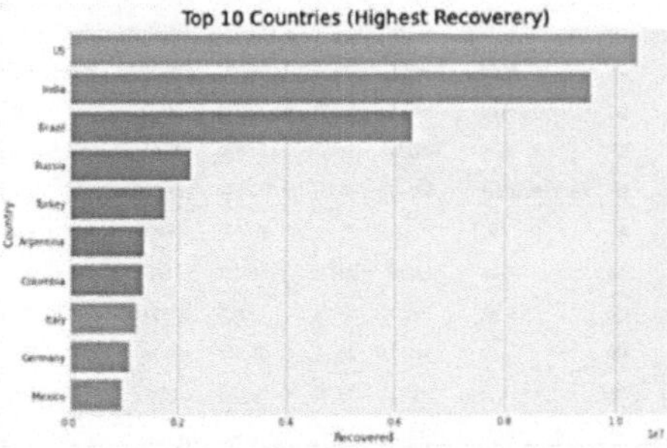

	Country	Confirmed	Deaths	Recovered	Active	Incident_Rate
177	US	17472522	313764	10397829	7176107	5303.280761
79	India	10004599	145136	9550712	308751	724.968638
23	Brazil	7162978	185650	6322955	654373	3369.871056
141	Russia	2792615	49744	2233060	509811	1913.609027
176	Turkey	1982090	17610	1753552	210928	2350.144566
6	Argentina	1531374	41672	1356755	132947	3388.312143
37	Colombia	1482072	40019	1354021	88032	2912.712259
85	Italy	1921778	67894	1226086	627798	3178.498010
66	Germany	1486839	25825	1096021	364993	1774.610876
114	Mexico	1301546	117249	962397	221900	1018.485576

FIGURE 8.11 Top ten countries with respect to Recovery cases as on December 19, 2020.

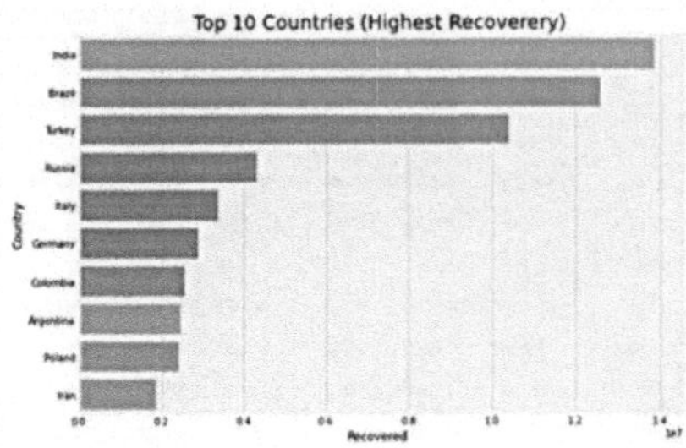

	Country	Confirmed	Deaths	Recovered	Active	Incident_Rate
79	India	16610481	189544	13867997	2552940	1203.654219
23	Brazil	14237078	386416	12593957	1256705	6697.928860
177	Turkey	4591416	38011	10397829	7176107	5443.996671
142	Russia	4699988	106108	4330610	263270	3220.615611
85	Italy	3949517	119021	3369048	461448	6532.248744
66	Germany	3276871	81542	2892350	302979	3911.096571
37	Colombia	2740544	70446	2555619	114479	5385.984018
6	Argentina	2824652	61176	2474514	288962	6249.814004
138	Poland	2751632	65222	2422659	263751	7270.485688
81	Iran	2377039	69120	1863405	444514	2830.045754

FIGURE 8.12 Top ten countries with respect to Recovery cases as on April 24, 2021.

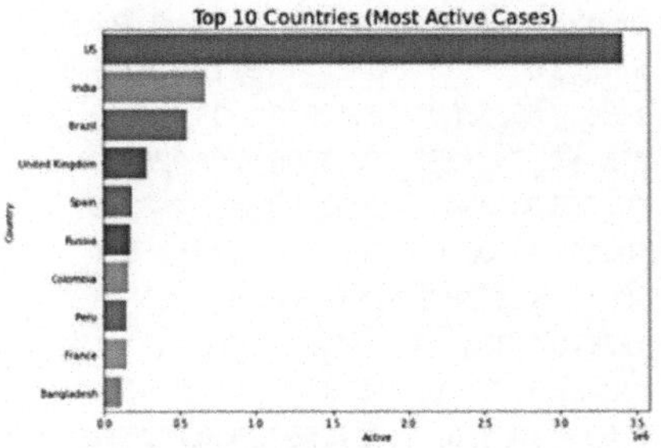

	Country	Confirmed	Deaths	Recovered	Active	Incident_Rate
174	US	5452115	170824	1865580	3415723	1654.832461
79	India	2702742	51797	1977779	673166	195.850247
23	Brazil	3359570	108536	2699080	551954	1580.532245
178	United Kingdom	322160	41466	1492	279202	474.560264
158	Spain	364196	28670	150376	185150	778.949183
140	Russia	930276	15836	741045	173395	637.461502
37	Colombia	476660	15372	301525	159763	936.778662
134	Peru	541493	26481	370717	144295	1642.289000
62	France	256534	30434	84214	141886	393.013938
13	Bangladesh	282344	3740	162825	115779	171.440317

FIGURE 8.13 Top ten countries with respect to active cases as on August 17, 2020.

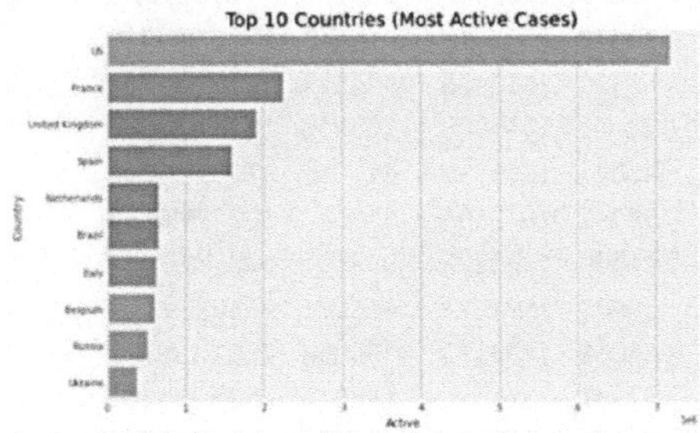

	Country	Confirmed	Deaths	Recovered	Active	Incident_Rate
177	US	17472522	313764	10397829	7176107	5303.280761
62	France	2499739	60347	188488	2250559	3829.637664
181	United Kingdom	1983017	66643	4340	1912034	2921.098434
161	Spain	1797236	48926	150376	1597934	3843.961804
123	Netherlands	674967	10492	8461	656014	3939.142123
23	Brazil	7162978	185650	6322955	654373	3369.871056
85	Italy	1921778	67894	1226086	627798	3178.498010
16	Belgium	621039	18455	0	602584	5358.581337
141	Russia	2792615	49744	2233060	509811	1913.609027
179	Ukraine	982937	17072	596218	369647	2247.547484

FIGURE 8.14 Top ten countries with respect to active cases as on December 19, 2020.

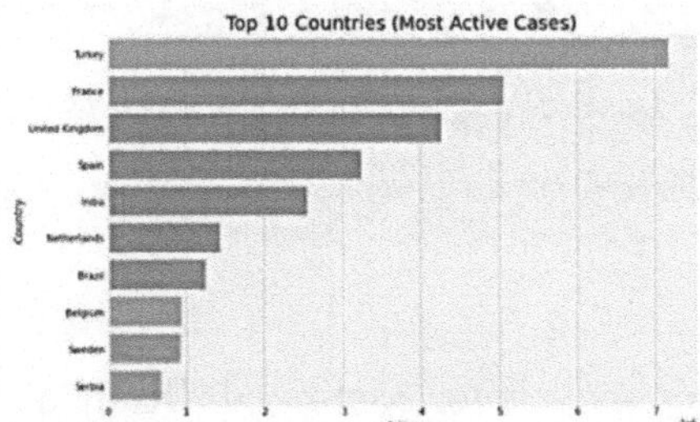

	Country	Confirmed	Deaths	Recovered	Active	Incident_Rate
79	India	16610481	189544	13867997	2552940	1203.654219
23	Brazil	14237078	386416	12593957	1256705	6697.928860
177	Turkey	4591416	38011	10397829	7176107	5443.996671
142	Russia	4699988	106108	4330610	263270	3220.615611
85	Italy	3949517	119021	3369048	461448	6532.248744
66	Germany	3276871	81542	2892350	302979	3911.096571
37	Colombia	2740544	70446	2555619	114479	5385.984018
6	Argentina	2824652	61176	2474514	288962	6249.814004
138	Poland	2751632	65222	2422659	263751	7270.485688
81	Iran	2377039	69120	1863405	444514	2830.045754

FIGURE 8.15 Top ten countries with respect to active cases as on April 24, 2021.

At the same time, we can see that patients recovered also increased rapidly, thus showing that patients are responding to the provided medication. Figure 8.16 shows the current situation of each country. Here, a number of confirmed, dead, recovered, and active cases can be checked for any country around the globe. This data can also be used to understand the magnitude of COVID-19 damage to a particular country. It also specifies which country needs a vaccine urgently.

	Country	Confirmed	Deaths	Recovered	Active	Incident_Rate
178	US	32000794	571285	0	0	9712.919243
79	India	16610481	189544	13867997	2552940	1203.654219
23	Brazil	14237078	386416	12593957	1256705	6697.928860
62	France	5502171	102656	336234	5063134	8429.408548
142	Russia	4699988	106108	4330610	263270	3220.615611
177	Turkey	4591416	38011	10397829	7176107	5443.996671
182	United Kingdom	4418709	127670	14388	4276651	6509.013257
85	Italy	3949517	119021	3369048	461448	6532.248744
162	Spain	3468617	77591	150376	3240650	7418.742592
66	Germany	3276871	81542	2892350	302979	3911.096571
6	Argentina	2824652	61176	2474514	288962	6249.814004
138	Poland	2751632	65222	2422659	263751	7270.485688
37	Colombia	2740544	70446	2555619	114479	5385.984018
81	Iran	2377039	69120	1863405	444514	2830.045754
114	Mexico	2323430	214504	1845008	263918	1818.130087
180	Ukraine	2071142	43884	1615812	411446	4735.796893
136	Peru	1745655	59012	1670805	15838	5294.380545
80	Indonesia	1636792	44500	1492322	99970	598.409744
46	Czechia	1618068	28920	1524041	65107	15109.447378
160	South Africa	1572985	54066	1499110	19809	2652.199872
124	Netherlands	1478097	17267	22714	1438116	8626.250104
32	Canada	1174977	23886	1067961	83130	3103.830963
35	Chile	1155902	25641	1084831	45430	6046.711458
141	Romania	1044722	27267	968746	48709	5430.602294
82	Iraq	1025288	15217	897966	112105	2549.040770
137	Philippines	989380	16674	883221	89485	902.874798
16	Belgium	968107	23954	0	944153	8353.227579

FIGURE 8.16 Current situation in all countries as on April 24, 2021.

8.4 PREDICTION MODEL

In this section, we have made prediction models for confirmed, Death, and Recovery cases. We have used Linear Regression, K-Nearest Neighbor Regressor, and Gradient Boosting Regressor for the prediction model and we compared the performances of these.

8.5 RESULT AND DISCUSSION

TABLE 8.4 Prediction Report of Confirmed Cases.

Classifier	R-squared	MSE
Linear regression	0.85069776665308	1848947755577.408
K-nearest neighbor regressor	0.9999537197886706	573130695.6825397
Gradient boosting regressor	0.9998159100727239	2279755970.3667355

TABLE 8.5 Prediction Report of Death Cases.

Classifier	R-squared	MSE
Linear regression	0.8953182038566116	280280200.9494639
K-nearest neighbor regressor	0.9998869946555886	302566.07936507935
Gradient boosting regressor	0.9998233504324985	472970.2594054472

TABLE 8.6 Prediction Report of Recovery Cases.

Classifier	R-squared	MSE
Linear regression	0.8133186916833616	2065557522265.441
K-nearest neighbor regressor	0.9999631049225796	408229968.96428573
Gradient boosting regressor	0.9998255745341253	1929951297.0042312

The table above shows the R-squared and mean squared error (MSE) values for all cases using the K-nearest neighbor regressor and the gradient boosting regressor. The coefficient of determination values are very good, 0.99 in all cases. MSE scores are also very good, as in this current study considering the long duration of this pandemic.

8.6 CONCLUSION

In this current work, we used the K-Nearest Neighbor regressor and the Gradient Boosting regressor to build predictive models for confirmed cases, recovery, and death, or mortality. The model performance is very good at predicting all cases. The R-squared value is very close to 1. The future directions of this work will analyze the steps taken by different countries in their policy measures. This kind of analysis can help other countries change their policies to combat this kind of pandemic.

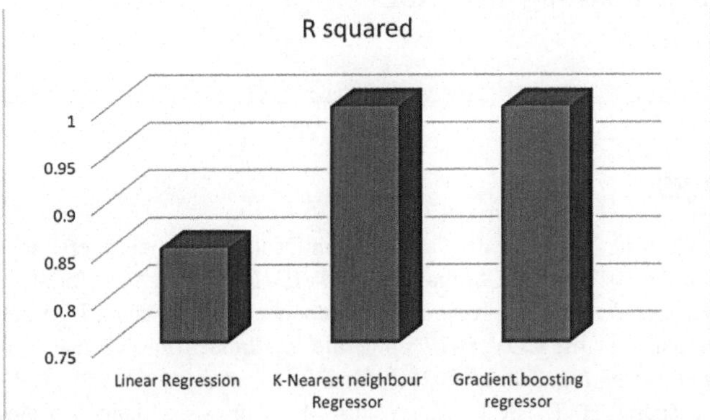

FIGURE 8.17 R-squared value.

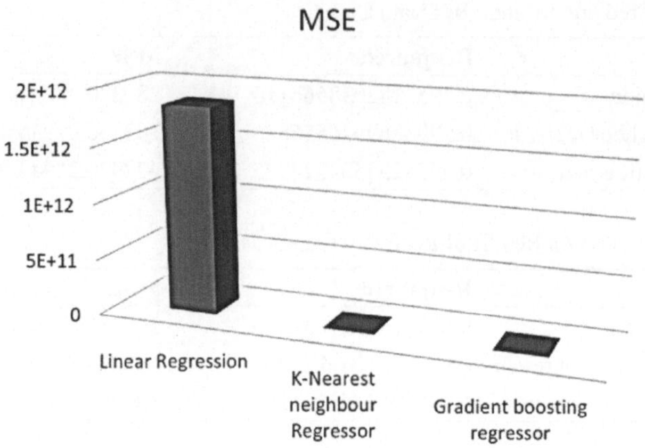

FIGURE 8.18 MSE value.

CONFLICT OF INTEREST

On behalf of all authors, the corresponding author states that there is no conflict of interest.

KEYWORDS

- **prediction**
- **COVID-19**
- **linear regression**
- **K-nearest neighbor regressor**
- **gradient boosting regressor**
- **R-squared**

REFERENCES

1. Sohrabi, C.; et al. World Health Organization Declares Global Emergency: A Review of the 2019 Novel Coronavirus (COVID-19). *Int. J. Surg.* **2020,** *76*, 71–76.
2. Roosa, K.; Lee, Y.; Luo, R.; Kirpich, A.; Rothenberg, R.; Hyman, J. M.; et al. Real-Time Forecasts of the COVID-19 Epidemic in China from February 5th to February 24th, 2020. *Infect. Dis. Model.* **2020,** *5*, 256–263
3. Yan, L. ; Zhang, H. T. ; Xiao, Y. ; Wang, M. ; Sun, C. ; Liang, J. ; et al. Prediction of Criticality in Patients with Severe COVID-19 Infection Using Three Clinical

Features: A Machine Learning- Based Prognostic Model with Clinical Data in Wuhan, 2020. medRxiv 2020.02.27.20028027

4. Stoecklin, S. B.; Rolland, P.; Silue, Y.; Mailles, A.; Campese, C.; Simondon, A.; et al. First Cases of Coronavirus Disease 2019 (COVID-19) in France: Surveillance, Investigations and Control Measures, January 2020. *Eurosurveillance* **2020,** *5* (6), 2000094.
5. Banerjee, A.; et al. Predicting the Recombination Potential of Severe Acute Respiratory Syndrome Coronavirus 2 and Middle East respiratory Syndrome Coronavirus. *J. Gen. Virol.* **2020,** *101* (12), 1251.
6. Coronavirus. World Health Organization [Online] 2020. https://www.who.int/healthtopics/coronavirus
7. Hui, D. S.; et al. Human cCronavirus Infections—Severe Acute Respiratory Syndrome (SARS), Middle East respiratory syndrome (MERS), and SARS- CoV- 2 . *Ref. Module Biomed. Sci.* **2020.**
8. McKeever, A. Here's What Coronavirus does to the Body [Online]. National Geographic:
9. https://www.nationalgeographic.com/science/2020/02/here-is-what- coronavirus-doesto-the- body/, 2020.16
10. Jaiswal, A. K.; Tiwari, P.; Kumar, S.; Gupta, D.; Khanna, A.; Rodrigues, J. J. Identifying Pneumonia in Chest X-rays: A Deep Learning Approach. *Measurement* **2019,** *145*, 511–518.
11. COVID-19 Coronavirus Pandemic [Online] 2020. worldometer: https://www.worldometers.info/ coronavirus/
12. Yadav, R. S. Data Analysis of COVID-2019 Epidemic Using Machine Learning Methods: A Case Study of India. *Int. J. Inf. Technol.* **2020,** *12*, 1321–1330.
13. Dey, S. K.; et al. Analyzing the Epidemiological Outbreak of COVID-19: A Visual Exploratory Data Analysis Approach. *J. Med. Virol.* **2020,** *92* (6), 632–638.
14. Mishra, N.; Majumder, A.; Rakshit, P. Design and Analysis of the Epidemic Novel Corona Virus. *J. Phys. Conf. Ser.* **2021,** *1797*, 012003.
15. Khan, A. I.; Shah, J. L.; Bhat, M. M. CoroNet: A Deep Neural Network for Detection and Diagnosis of COVID-19 from Chest X-Ray Images. *Comput. Methods Programs Biomed.* **2020,** *196*, 105581.
16. Wang, L.; Zhong, Q. L.; Wong, A. Covid-net: A Tailored Deep Convolutional Neural Network Design for Detection of Covid-19 Cases from Chest X-Ray Images. *Sci. Rep.* **2020,** *10* (1), 1–12.
17. Narin, A.; Kaya, C.; Pamuk, Z. Automatic Detection of Coronavirus Disease (COVID- 19) Using X-ray Images and Deep Convolutional Neural Networks, 2020. arXiv:2003.10849v1
18. Shen, D.; Wu, G.; Suk, H. I. Deep Learning in Medical Image Analysis. *Ann. Rev. Biomed. Eng.* **2017,** *19*, 221–248.
19. Rajpurkar, P.; et al. Chexnet: Radiologist-Level Pneumonia Detection on Chest X-rays with Deep Learning, 2017. arXiv preprint arXiv:1711.05225
20. Wang, H.; Yong, X. Chestnet: A Deep Neural Network for Classification of Thoracic Diseases on Chest Radiography, 2018. arXiv preprint arXiv:1807.03058 (2018).17
21. Hemdan, E. E. D.; Shouman, M. A.; Karar, M. E. Covidx-Net: A Framework of Deep Learning Classifiers to Diagnose Covid-19 in X-Ray Images, 2020. arXiv preprint arXiv:2003.11055
22. Ayan, E.; Unver, H. M. Diagnosis of Pneumonia from Chest X-Ray Images Using Deep Learning. Scientific Meeting on Electrical-Electronics Biomedical Engineering

and Computer Science (EBBT), Istanbul, Turkey, 2019; pp 1–5. https://doi.org/10.1109/EBBT.2019.8741582

23. Gaal, G.; Maga, B.; Luk ´ acs, A. Attention U-Net Based Adversarial Architectures for Chest X-ray Lung Segmentation, 2020. arXiv:2003.10304
24. Zhang, J.; Xie, Y.; Li, Y.; Shen, C.; Xia, Y. COVID-19 Screening on Chest X-ray Images Using Deep Learning based Anomaly Detection, 2020. arXiv:2003.12338v1
25. Liang, G.; Zheng, L. A Transfer Learning Method with Deep Residual Network for Pediatric Pneumonia Diagnosis. *Comput. Methods Programs Biomed.* **2020,** *187*, 104964.
26. Jaiswal, A.; Gianchandani, N.; Singh, D.; Kumar, V.; Kaur, M. Classification of the COVID- 19 Infected Patients Using DenseNet201 Based Deep Transfer Learning. *J. Biomol. Struct. Dyn.* **2020,** 1–8. https://doi.org/10.1080/07391102.2020.1788642
27. Ucar, F.; Korkmaz, D. COVIDiagnosis-Net: Deep Bayes-SqueezeNet Based Diagnosis of the Coronavirus Disease 2019 (COVID-19) from X-ray Images. *Med. Hypotheses* **2020,** *140*, 109761.
28. Sahinbas, K.; Catak, F. O. Transfer Learning Based Convolutional Neural Network for COVID19 Detection with X-Ray Images [Online] 2020. https://www.ozgurcatak.org/files/papers/covid19- deep-learning.pdf
29. Das, N. N.; Kumar, N.; Kaur, M.; Kumar, V.; Singh, D. Automated Deep Transfer
30. Learning- Based Approach for Detection of COVID-19 Infection in Chest X-rays. *IRBM* **2022,** *43* (2), 114–119. https://doi.org/10.1016/j.irbm.2020.07.001, 2020.18
31. Apostolopoulos, I. D.; Mpesiana, T. A. Covid-19: Automatic Detection from X-Ray Images Utilizing Transfer Learning with Convolutional Neural Networks. *Phys. Eng. Sci. Med.* **2020,** *43*, 635–640.
32. Singh, D.; Kumar, V.; Kaur, M. Classification of COVID-19 Patients from Chest CT Images Using Multi-Objective Differential Evolution-Based Convolutional Neural Networks. *Eur. J. Clin. Microbiol. Infect. Dis.* **2020,** *39*, 1379–1389. https://doi.org/10.1007/s10096-020-03901-z

CHAPTER 9

Mucormycosis in India: Pathways of Spread in Post-COVID-19 and Diabetic Patients

NITIN ARORA[1], ADITYA KAKDE[2], GARIMA PANDEY[3], HUMA NAAZ[2], LAKSHAY GOYAL[2], SHASHWAT TIWARI[2], and AHMED A. ELNGAR[4]

[1]*Electronics & Computer Discipline, Indian Institute of Technology, Roorkee, India*

[2]*School of Computer Science, University of Petroleum & Energy Studies, Dehradun, India*

[3]*Department of Computer Science, National Institute of Technology Karnataka Surathkal, India*

[4]*Faculty of Computers and Artificial Intelligence, Beni-Suef University, Beni-Suef City, Egypt*

ABSTRACT

Mucormycosis, renowned as black fungus, is an uncommon and invasive fungal infection that commonly targets immunocompromised patients. A group of molds called Mucor causes mucormycosis. According to the Centre for Disease Control and Prevention (CDC), this type of infection has a mortality rate of 46%. According to a survey, mucormycosis has affected only 109 thousand people in India annually in the past which is 1% of the total population. But recently, the cases are increasing at a very high pace. It's presumed that uncontrolled diabetes and careless use of steroids are two main factors for the spread of mucormycosis in India. This fungus spread through the paranasal sinus into orbit affecting, the eyes,

Global Digital Transformation and the Covid-19 Pandemic.
Rajdeep Chowdhury, K. Martin Sagayam, Neyara Radwan, Ahmad A. Elngar (Eds.)

brain, and lungs. Therefore, it is vital to diagnose mucormycosis at the right point in time for timely treatment. Hence, this paper provides insight into mucormycosis occurrence in diabetic and COVID-19 patients and covers the types of disease (mucormycosis/COVID-19), and investigates how these molds are getting a chance to invade the body of immunocompromised patients. In addition, this paper addresses possible treatments and preventive measures.

9.1 INTRODUCTION

Black fungus, biologically known as mucormycosis, is a severe and rare fungal infection caused by a group of molds known as Mucor.[1] It is termed "black" because it looks like black when it affects the skin.

Now talking about Mucor, it is omnipresent and can be found in soil, manure, decaying bread, fruits, and vegetables. Sense organs ears, skin, tongue, eyes, and nose are the mediums through which this infection can invade the body.[1] Humans' immune system is strong enough to deal with it on its own, but if the immune system is breached by another illness/disease, like any other fungi, these Mucor can easily invade the human tissues.

People suffering from COVID-19, HIV/AIDS, and other viral diseases, congenital bone marrow disease, severe burns, cancers, and untreated or irregularly treated diabetes have reduced immunity, and are more prone to developing mucormycosis.[2] A novel virus named COVID-19 emerged from the part of the coronavirus family that came into existence at the time of transmission from a bat to a person.[3,4] Nowadays, most of the cases are of COVID-19 patients. The patients suffering from this virus consumed steroids as a treatment. Consumption of steroids increases the sugar level in the blood, resulting in decreased immunity. With the registered cases, it was seen that Rhino cerebral, a type of mucormycosis, is causing this fungal infection.

Rhino cerebral mucormycosis, also known as Zygomycosis, is a rare opportunistic disease affecting the nose, paranasal sinuses, and brain. Rhino cerebral mucormycosis generally affects people with diabetes and those in immune compromised states. The pathogens that cause Rhino cerebral mucormycosis are prevalent but may be more likely to cause infection in moist, temperate climates. Exposure to Mucoromycete molds causes it. These organisms can be found in leaves, soil, compost piles, and rotting wood.[5] Mucormycosis is contracted by inhaling mould spores that have been exposed to the air. A sinus (pulmonary) exposure is the term used to describe this situation. You could have an infection in your eyes, lungs, face, sinuses, or central nervous system if you take antibiotics (rarer). A wound or burn can potentially expose you to the fungus. In such situations, the wound or burn becomes the source of infection. While these moulds may occur naturally in the environment, not everyone who is exposed to them will get a fungal illness. If you have a compromised immune system, you may be more susceptible to this form of infection. Burns, scratches and scrapes, cancer, a recent organ transplant, HIV or AIDS, diabetes, and surgery are all conditions that can raise your risk.[5] The prognosis of mucormycosis infections ranges from fair to poor; there is about a 50% death rate that grows to about 85% for Rhino cerebral and GI infections.

But this fungus cannot be only considered vulnerable to a COVID-19 patient, as it can also affect an immune competent state person. The fungus can invade the human body through various mediums like the consumption of bread containing molds. The number of black fungus cases is gradually increasing in India, and according to conditions,[6] it may rise in the upcoming weeks. Figure 9.1 shows the number of instances of mucormycosis in some of the states in India.

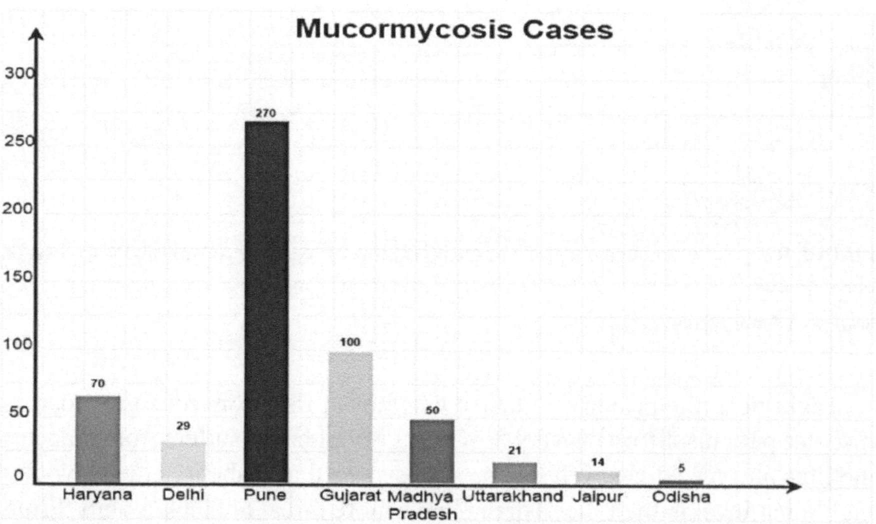

FIGURE 9.1 Recent mucormycosis cases in diverse cities and states of India.[7–15] The x-axis shows states in India, and the y-axis indicates the number of cases of mucormycosis till the paper submission.

9.2 CAN THERE BE A RISK IF A PERSON HAS DIABETES?

Yes, the diabetes rate is shown in Figure 9.2, and an individual can make the necessary decision based on this information. If a person's diabetes rate lies between 100 and 125 mg/dL, they are at mild risk. But, if the diabetes rate is above 126 mg/dL, then they are at high risk, but in both cases, it's essential to see a doctor as sugar level in the blood is inversely proportional to the immune system.

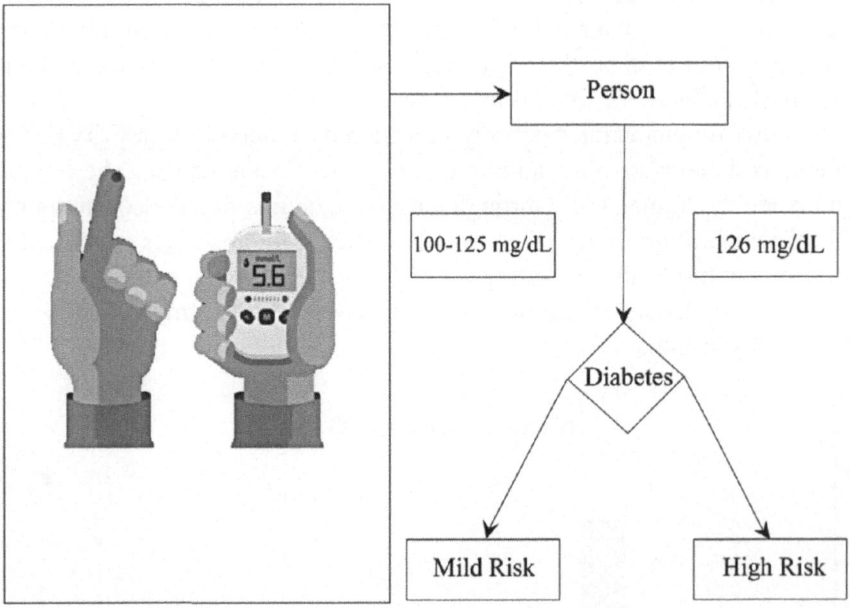

FIGURE 9.2 Level of sugar (mg/dL) in the blood can increase the risk of mucormycosis or black fungus.

Source: Adapted from Ref. [16]

Right now, the primary audience which is at risk from mucormycosis is diabetes patients. More than 8.9% of India's population suffers from diabetes mellitus and recent cases. It has been observed that diabetic patients with a high sugar level in their bloodstream have more threat of black fungus. From Figure 9.3, it can be analyzed that the spread of this rare disease in diabetic patients can increase in the upcoming days. The average was computed using data from the 2017–2018 years, including patients aged 45, 45–59, 60, and above.

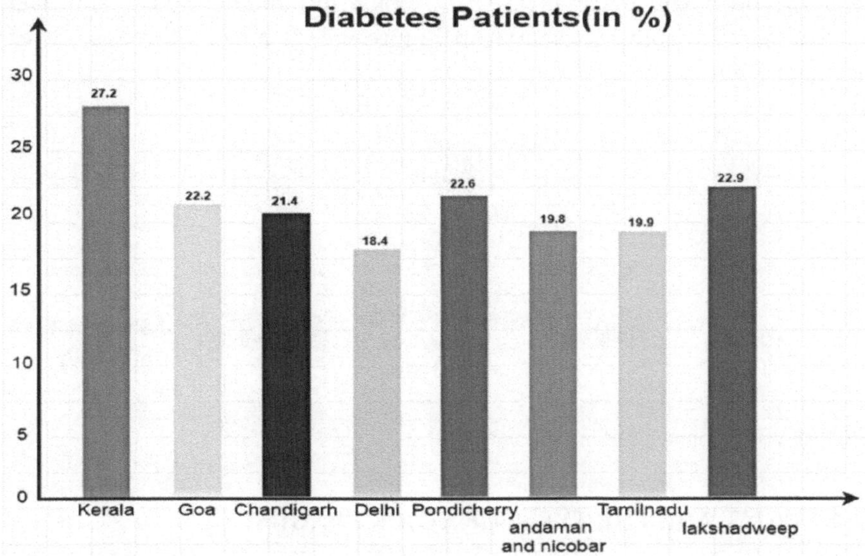

FIGURE 9.3 Top eight states with highest diabetes patients (in %). The x-axis indicates the states of India, and the y-axis shows the total number of cases in diabetes patients.

Source: Adapted from Ref. [17]

9.3 WHAT IS THE DIFFERENCE BETWEEN NORMAL FUNGI AND BLACK FUNGI?

It is essential to know if a person has a normal or a black fungus as the symptoms may be the same, and the wrong medication can lead to side effects. A slight difference is presented in Table 9.1.

TABLE 9.1 Presents How the Common Fungus Differs From the Black Fungus.

Normal fungus	Black fungus
It usually occurs due to sweat	Usually occurs due to low immunity and inhalation of molds
Faded skin color	Black in color
Occurrence is common	Occurrence is rare

9.4 WHAT ARE THE COMMON SYMPTOMS OF BLACK FUNGUS?

Common symptoms may be fever, headache, reddish, and swollen skin near the nose and eyes, facial pain, cough-producing fluids, and even shortening of breath,[1] as shown in Figure 9.4. It attacks the nose, eyes, brain, and lungs.

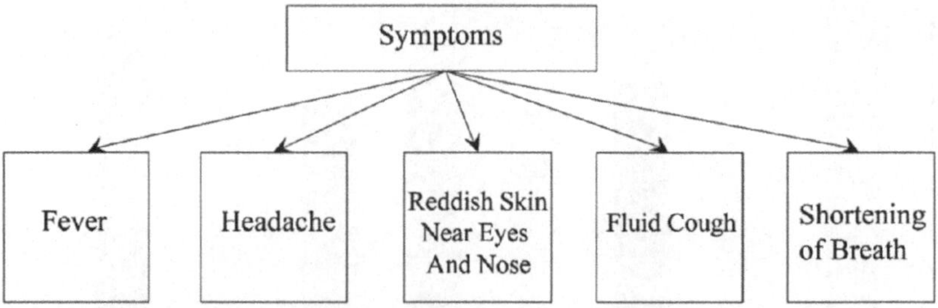

FIGURE 9.4 Symptoms of black fungus.

9.5 WHAT ARE THE TYPES OF BLACK FUNGUS?

There are six types of black fungus identified until now, where rhino cerebral mucormycosis is the one mainly targeting decreased immune patients. Figure 9.5 shows the first two types of mucormycosis: rhinocerebral, which affects the brain and eyes, and the second in pulmonary, which affects the lungs.

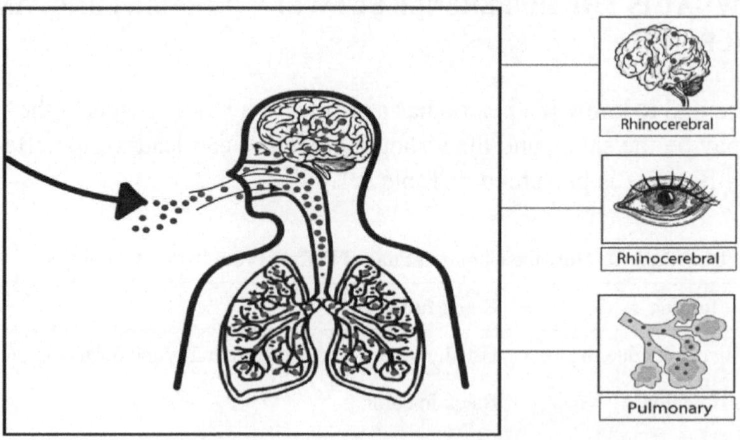

FIGURE 9.5 Mucormycosis or black fungus forms that are most commonly encountered.

- **Rhinocerebral Mucormycosis**[18]**:** Also known as Rhino-orbital cerebral, it can affect the face and affect the nose, eyes, or brain. In this disease, the infection can spread to the eye orbit, nose, and inside the brain leading to loss of vision. Symptoms of this disease include headache, nose block, and bleeding nose. If observing, facial symptoms include skin discoloration, numbness, pain, and one-sided face swelling.
- **Pulmonary Mucormycosis**[19]: Pulmonary mucormycosis is an infrequent fungal infection due to lymphatic spread. Some disseminated cancer cells develop the capability to enter the membranes of the lymph vessels, then move to the other spots of the body via the bloodstream, termed as a lymphatic spread or hematogenous spread, or by the inhaling of spores. It is arduous to diagnose the disease in an early stage and has no effective cure. The Mucorales enter the respiratory passage further, entering the veins, lymphatics, and arteries, causing infarction (tissue death due to shortage of blood supply to the concerned spot) and thrombosis (blood clot in the blood vessel). These both are critical complications and can be deadly. The individuals more prone to this disease are the ones with a less defensive immune system, may have diabetes, may have an organ or stem cell transplant, or have hematologic malignancies (these are cancers that attack the bone marrow, blood, and lymph nodes). Fever, cough, chest pain, shortness of breath, and blood in the cough are all symptoms of pulmonary (lung) mucormycosis. The above types are marked as having the highest incidence. More kinds of mucormycosis are shown in Figure 9.6. The gastrointestinal occurs in the stomach, cutaneous occurs in the skin, and disseminated passes through the bloodstream and affects the brain, spleen, skin, and heart. The last one is sinonasal, which affects the respiratory/nasal part.

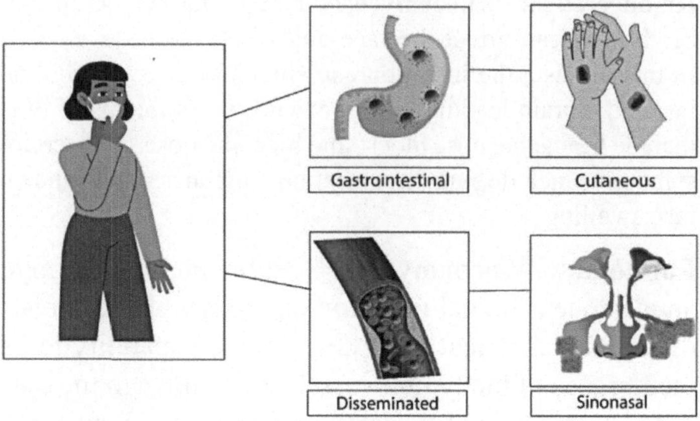

FIGURE 9.6 Four more types of mucormycosis or black fungus.

- **Gastrointestinal Mucormycosis**[20]**:** It is the most uncommon deadly disease, with a percentage of 4–7% of all the cases. Malnutrition, particularly in premature neonates, newborns, and young adults, is a condition that puts a person at risk of disorder or disease. Ingestion is the most common way for fungi to penetrate the gastrointestinal tract. The stomach, followed by the large intestine and ileum, is the most common spot of gastrointestinal mucormycosis. Abdominal pain, nausea, vomiting, and gastrointestinal bleeding are some of the symptoms. Fever and hematochezia are also potential side effects.
- **Cutaneous Mucormycosis**[21]**:** The causative agents of cutaneous mucormycosis are omnipresent and enter the skin by straight inoculation because of several natures of lesions. These comprise animal bites, needle sticks, traffic accidents, cuts due to burns, or natural disasters. This form of mucormycosis can even be seen in individuals that do not have a weak immune system (immune competent). Blisters or ulcers, infected areas can turn black, extreme redness, swelling around a wound, and pain is some skin symptoms.
- **Disseminated Mucormycosis**[22]**:** The individuals that are more prone to this form of mucormycosis are individuals with a low count of neutrophils, transplants, and hematologic malignancies. Its causative agents can also be seen in patients who are on deferral treatment. This disease occurs when the bloodstream reaches the other body parts and attacks them. The brain is most typically affected by

this infection. The other body organs affected are the heart, spleen, and skin. Its symptom is coma in the brain.

- **Sinonasal Mucormycosis**[23]**:** This is a severe, rare, opportunistic, and life-threatening fungal infection that happens primarily in patients with low defensive immune capability. Asinonasal fungal infection that occurs from fungi infects the respiratory/nasal mucosa the most. It is categorized into two types:
- **Invasive:** The existence of fungal hyphae inside the nasal mucosa.
- **Non-invasive:** The fungal hyphae are not present inside the nasal mucosa in this state. Its symptoms include nasal discharge, dark ulcers within the nasal canal, or dark ulcers on the roof of the mouth.

The summary of the above data is presented in Table 9.2.

TABLE 9.2 A Summary of the Types of Black Fungus.

Sr. No.	Type	Effects on	Symptoms
1	Rhino-orbital cerebral/rhino cerebral mucormycosis	Face, nose, eyes, or brain	Headache, nose block, bleeding nose, skin discoloration, numbness, and pain, one-sided face swelling.
2	Pulmonary mucormycosis	Lungs	Fever, cough, chest pain, shortness of breath, blood in cough.
3	Gastrointestinal mucormycosis	Stomach, followed by the large intestine and ileum	Abdominal pain, nausea, vomiting, and gastrointestinal bleeding, fever, and hematochezia.
4	Cutaneous mucormycosis	Skin	Blisters or ulcers, infected area can turn black, extreme redness, swelling around a wound, and pain.
5	Disseminated mucormycosis	From bloodstream to brain, heart, spleen, and skin	Coma in the brain.
6	Sinonasal mucormycosis	Nasal cavity	Nasal discharge, dark ulcers within the nasal canal.

9.6 WHY MOST THE COVID-19 PATIENTS ARE GETTING AFFECTED BYMUCORMYCOSIS OR BLACK FUNGUS?

According to one study, earlier mucormycosis only affected 109 thousand people per year, 1% of the total population in India. But recently, the cases are increasing at a high pace. The reason for these increasing mucormycosis cases is the devastating ongoing second wave of Corona. The patients suffering from COVID-19 are at higher risk because of their weak immune systems. These patients are contributing to the rise in the number of cases of mucormycosis. Public health experts also blame steroids in COVID-19 treatments, as it increases blood sugar levels and affects the lungs to an extent.[24]

According to the International Diabetes Foundation (IDF) 2019 report, 463 million people live with diabetes, which is likely to increase to 700 million by the year 2045, as stated by public health experts. "Up to 75% of mucormycosis cases occur among COVID-19 patients with diabetes." Therefore, it could be a possible explanation for the influx of mucormycosis cases in India. As discussed, high doses of steroids could be a possible reason for the increased number of mucormycosis patients in India to control COVID-19. Steroids increase blood sugar and mimic a diabetes situation.[25]

The more general symptoms of mucormycosis include unspecific runny nose, fever, headache, and one-sided facial swelling. But then it might turn to some specific Symptoms such as black lesions on the nasal bridge or upper inside of the mouth or necrotic skin that would look black, and this is why it is also known as black fungal disease. Mucormycosis mainly destroys the tissue from where it starts spreading and could ruin the neighboring organs' tissues; it could be the eye or other organ. Therefore, to prevent the infection from spreading to the brain, some patients have undergone eye surgery.[26] Mucormycosis has typically been seen after 2 to 3 weeks of COVID-19 infection in all recovered or recovering patients. Immune-suppressing steroids prescribed during the treatment of COVID-19 can increase blood sugar levels and leave people vulnerable to black fungal infections.

9.7 HOW SHOULD A PERSON KNOW IF HE/SHE IS SUFFERING FROM COVID-19 OR BLACK FUNGUS AS MOST OF THE SYMPTOMS ARE THE SAME?

Figure 9.7 guides how a COVID-19 patient should identify these particular diseases if a patient has those symptoms like fever, headache, and shortening of breath.

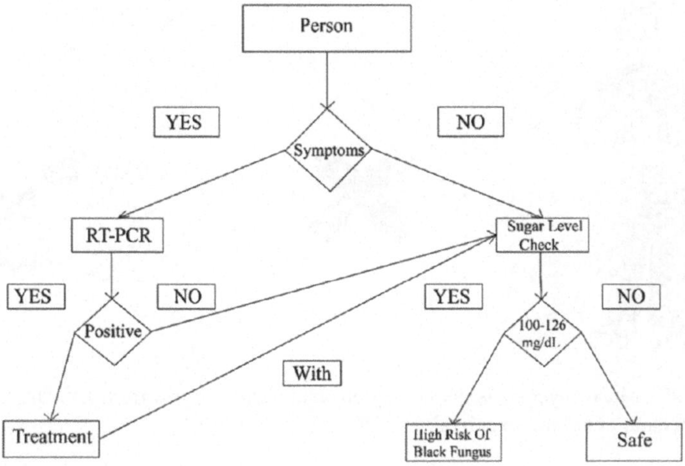

FIGURE 9.7 Strategy to know a particular disease with typical symptoms.

From Figure 9.7, if a person has typical symptoms, then first, he/she should get the RT-PCR test still, if it does not mean there are no specific symptoms. Still, some rise is seen like reddish skin near eyes and nose or fluid cough then one should go for sugar level check as the person redness can also be seen during conjunctivitis, or even reddish nose can be seen in some people who have sneezing problem. If the RT-PCR test came positive, then treatment needs to be done. After that still, there is a need for sugar level tests as it was found that all medicines contain a high number of steroids which increases the sugar level in the blood. If the RT-PCR test came negative, there is still a need for a sugar level test because the black fungus mentioned previously has some common symptoms. If the sugar level came between 100 and 126 gm/dL, then immediate treatment is required, and if it's not, then a person is safe from black fungus, but one can have COVID-19 or a common fever, headache, etc. Also, repeat the sugar level check after 2–3 weeks of COVID-19 infection.

9.8 IS MUCORMYCOSIS OR BLACK FUNGUS CONTAGIOUS?

Unlike COVID-19, it is not a contagious disease, as shown in Figure 9.8, but it can transfer from one to another if someone comes in contact with the scrubbed fungus part.Also, it is essential to make sure that it is not left untreated.[2]

FIGURE 9.8 Mucormycosis is not contagious and cannot spread from animals to humans, person to person, or animal to animal.

9.9 ANALYSIS AND DISCUSSION

The reviewed content reveals that black fungus is primarily infecting people through two gateways: Diabetes and COVID-19.

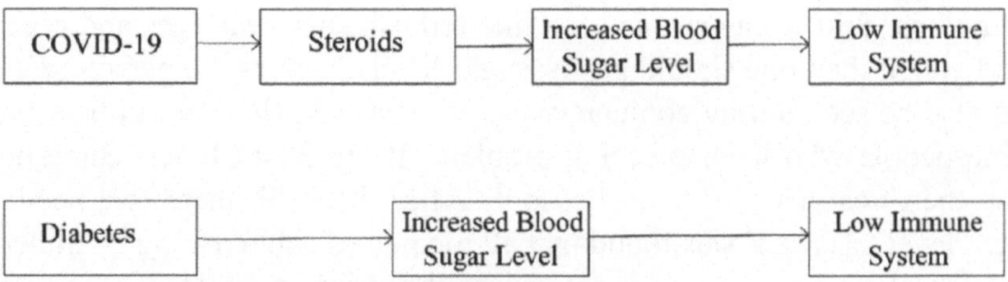

FIGURE 9.9 Various gateways pointing toward the same point.

From Figure 9.9, both the gateways result in the same data point which is decreased immune system. But how is a mold getting to know that if a person's immune system is weakened? How is mold getting a chance of entering the body?

First, we need to know where these molds are coming from? Saprophytic fungi feed on non-living organic matter to help plants process soil mineralization and carbon cycle.[27] Rhino cerebral mucormycosis being, a saprophytic of the class Phycomycetes,[28] triggers these molds to get into the air.[29]

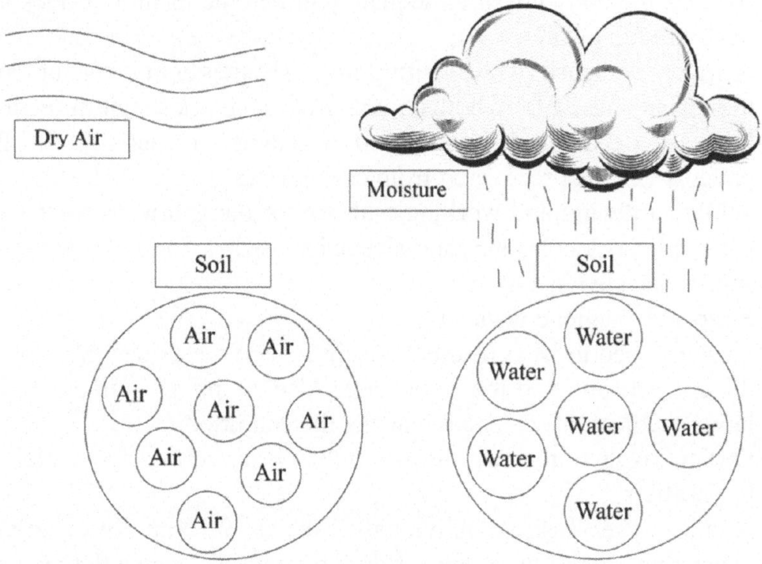

FIGURE 9.10 Water molecules replace air molecules in the soil.

As shown in Figure 9.10, during rain, the air molecules present in the soil are displaced by the water molecules, thus resulting in exposure of air from soil that consists of molds.

Dust comprises various particles like bacteria, pollen, decomposing insects, and molds.[30] Molds coming from air particles through soil only grow in moisture and cannot reside in dry areas. So, the molds won't be present in dust.[31]

Therefore, it can be said that if the room environment is moist, then there can be chances that the room's air molecules contain molds. These molds can invade anyone who can inhale air, and the recent studies[5] affected only those with low immune systems. Therefore, air, when inhaled by a person with a decreased immune system in a moist environment, can cause eye blindness, and brain and lung problems.

To keep the room environment dry, one should maintain the temperature of 24–30°C and increase fan speed to increase air movement.[32]

9.10 WHAT PREVENTION STEPS NEEDS TO BE TAKEN?[33]

- Control hyperglycaemia (high blood glucose/sugar is referred to as hyperglycaemia in medical terms). The blood sugar level increases because the body doesn't adequately utilize the insulin or does not have enough insulin.

- Always wear masks while visiting any dusty sites or moving outdoors to prevent oneself from both coronavirus and black fungus infection.
- Diabetes needs to be monitored constantly, essentially for the patients who just recovered from coronavirus.
- While gardening and working with soil or dung, always wear long pants, gloves, boots, and long-sleeved shirts to avoid mucormycosis spores detected in soil.
- Keep your hygiene in check.
- Steroids need to be consumed wisely for the prescribed duration at the correct time and the right dose. Also, stopping immunomodulating medications can help manage the condition.
- During oxygen remedy, utilize only clean, uninfected water in humidifiers.
- All the oxygen-assisted individuals in the hospital must be examined constantly around the nares for dark pigment. If even a dot appears around the nares, the treatment must begin instantly. The sole cure for this condition is antifungal vaccination.
- Consume antifungal/antibiotics wisely.
- Do not overlook warning symptoms and signs. Symptoms include nasal block, face swelling, numbness, reduced immunity, sinus pressure, toothache, and slackening of teeth.
- Do not waste critical time starting medication with mucormycosis.

9.11 WHAT ARE THE POSSIBLE TREATMENTS?[33]

- Supervised medical intervention and intravenous antifungal medications can treat this fungal infection.
- Elimination of affected tissues through surgery is also a possible treatment.
- Amphotericin B, an Antifungal drug, destroys the fungal cell membrane and is used widely to treat black fungal infection. According to a medical community, 4–6 vials of Amphotericin B are given to an infected patient per day for 10 days. The cost of a vial of Amphotericin B range from INR 3000 to INR 7000.
- Other antifungal drugs used for treatment are isavuconazole and posaconazole.
- Home remedies cannot treat mucormycosis.

9.12 CONCLUSION AND FUTURE WORK

From the above studies, it was seen that the recent growth of mucormycosis or black fungus is due to a low immune system which can be considered as the root cause as both the gateways discussed above lead to the same data point. The type of mucormycosis or black fungus, which has more occurrences is rhino cerebral, whose molds can affect the eyes and brain. The previous work showed that patients who are COVID-19 positive are at high risk due to consumption of steroids, which also leads to a decreased immune system. Analysis was also conducted, which showed that the problems are arising mainly from two gateways, namely, diabetes and COVID-19, but leads to the same data point, that is, an immune system, but didn't make it clear about how these molds of mucormycosis are communicating with the patients to get to know if the immune system is weak or not and, therefore, this paper shows how these molds are getting a chance to enter into the body. Further, preventions and treatments which are covered suggested that right now, the only solution to this situation is to remove the affected tissue. For Rhino cerebral, it is the removal of eyes so that the molds don't reach the brain.

Some doctors have said that the fungus is also infecting because of late hospitalization from the above data. Also, from the above analysis, the molds need to get into the body through some channel. Maybe the moist environment in the hospital results in molds living in air molecules, and because of that, it gets a chance to enter the body.

Further, four cases in Patna, Bihar, India, have been recently cited, which suggested that there has been a spread of White Fungus, which according to the doctors, is more dangerous than black fungus. It not only affects the eyes, brain, lungs, and skin but also affects each and every part of the body, but the reason causing this is the same: a weak immune system.

CONFLICT OF INTEREST

The authors whose names are listed immediately below certify that they have NO affiliations with or involvement in any organization or entity with any financial interest (such as honoraria; educational grants; participation in speakers' bureaus; membership, employment, consultancies, stock ownership, or other equity interest; and expert testimony or patent-licensing arrangements), or non-financial interest (such as personal or professional relationships, affiliations, knowledge or beliefs) in the subject matter or materials discussed in this manuscript.

KEYWORDS

- **mucormycosis**
- **black fungus**
- **COVID-19**
- **molds**
- **diabetic**
- **zygomycosis**

REFERENCES

1. Suryanarayanan, T. S. COVID-19 and Black Fungus: What Is Mucormycosis? [Online] 2021. https://science.thewire.in/the-sciences/covid-19-and-black-fungus-what-is-mucormycosis/ (accessed May 20, 2021)
2. Acharjee, S. Explained: What is black Fungus. [Online] 2021. https://www.indiatoday.in/india-today-insight/story/explained-what-is-black-fungus-1803263-2021-05-16 (accessed May 20, 2021)
3. Kakde, A.; Sharma, D.; Arora, N. Optimal Classification of COVID-19: A Transfer Learning Approach. *Int. J. Comput. Appl.* 2020, *975*, 8887.
4. Goyal, L.; Arora, N. Deep Transfer Learning Approach for Detection of Covid-19 from Chest X-Ray Images. *Int. J. Comput. Appl.* 2020, *975*, 8887.
5. Cherneyk. What is Mucormycosis. [Online] 2017. https://www.healthline.com/health/mucormycosis (accessed Oct 18, 2017)
6. Jazeera, A. L. India Sees 259,551 New COVID Cases, 'Black Fungus' Adds to Woes. [Online] 2021. https://www.aljazeera.com/news/2021/5/21/india-sees-259551-new-covid-cases-black-fungus-adds-to-woes (accessed May 22, 2021)
7. Singh, S. Gurugram After, Being a Hotspot of Coronavirus, Became A New Hotspot for Black Fungus. [Online] 2021. https://www.livehindustan.com/ncr/story-gurugram-is-also-becoming-a-hotspot-of-black-fungus-after-covid-19-4039248.html (accessed May 20, 2021)
8. Dutt, A. At Least 29 'Black Fungus' Cases in AIIMS, Ganga Ram Spark Worry. [Online] 2021. https://www.hindustantimes.com/cities/others/at-least-29-black-fungus-cases-in-aiims-ganga-ram-spark-worry-101620849405898.html (accessed May 21, 2021)

9. Press Trust of India. Covid-19: 270 Black Fungus Cases Reported in Pune; Therapy SOPs Framed. [Online] 2021. https://www.business-standard.com/article/current-affairs/covid-19-270-black-fungus-cases-reported-in-pune-therapy-sops-framed-121051400795_1.html (accessed May 17, 2021)
10. Express News Service. Ahmedabad, Vadodara, Gujarat: Over 100 Black Fungus Cases, Govt Announces separate Wards for Patients. [Online] 2021. https://indianexpress.com/article/cities/ahmedabad/gujarat-over-100-black-fungus-cases-govt-announces-separate-wards-for-patients-7307579/ (accessed May 19, 2021)
11. NEWS18 BHOPAL. 50 Black Fungus Cases Reported in MP, CM Chouhan Orders Special Treatment Wards. [Online] 2021. https://www.news18.com/news/india/black-fungus-cases-now-50-in-mp-cm-chouhan-orders-special-treatment-wards-3731201.html (accessed May 19, 2021)
12. Azad, S. Uttarakh and Records its First Black Fungus Death, 21 Cases in State so Far. [Online] 2021. https://timesofindia.indiatimes.com/city/dehradun/uttarakhand-records-its-first-black-fungus-death-21-cases-in-state-so-far/articleshow/82689167.cms (accessed May 19, 2021)
13. By HT Correspondent. Cases of Covid-Linked Mucormycosis Reported Across India. [Online] 2021. https://www.hindustantimes.com/india-news/cases-of-covid-linked-mucormycosis-reported-across-india-101620823722866.html (accessed May 19, 2021)
14. TNN. Four More Cases of Black Fungus in Odisha. [Online] 2021. https://timesofindia.indiatimes.com/city/bhubaneswar/four-more-cases-of-black-fungus-in-odisha/articleshow/82567732.cms (accessed May 15, 2021)
15. Pradhan, A. Bhubaneswar Reports First Black Fungus Case in Odisha. [Online] 2021. https://timesofindia.indiatimes.com/city/bhubaneswar/bhubaneswar-reports-first-black-fungus-case-in-odisha/articleshow/82543141.cms (accessed May 16, 2021)
16. Diagnosis. [Online] 2021. https://www.mayoclinic.org/diseases-conditions/diabetes/diagnosis-treatment/drc-20371451#:~:text=A%20blood%20sample%20will%20be%20taken%20after%20an%20overnight%20fast,separate%20tests%2C%20you%20have%20diabetes (accessed May 17, 2021)
17. Mishra, N. L.; Gaikwad, R.; Bapat, S. V. Diabetes among Senior Citizens More Prevalent in Urban India: LASI Report. [Online] 2021. https://www.downtoearth.org.in/news/health/diabetes-among-senior-citizens-more-prevalent-in-urban-india-lasi-report-75038 (accessed May 12, 2021)
18. Ajith Kumar, A. K.; Gupta, V. Rhino-Orbital Cerebral Mucormycosis. In *StatPearls* [Internet]; StatPearls Publishing: Treasure Island, FL, 2021. https://www.ncbi.nlm.nih.gov/books/NBK557429/ [updated Feb 8, 2021].
19. Agrawal, R.; Yeldandi, A.; Savas, H.; Parekh, N. D.; Lombardi, P. J.;Hart, E. M. Pulmonary Mucormycosis: Risk Factors, Radiologic Findings, and Pathologic Correlation. , *Radiographics* 2020, *40* (3), 656–666.
20. Goel, P.; Jain, V.; Sengar, M.; Mohta, A.; Das, P.; Bansal, P. Gastrointestinalmucormycosis: A Success Story and Appraisal of Concepts. *J. Infect. Public Health* 2013, *6* (1), 58–61. ISSN 1876-0341. https://doi.org/10.1016/j.jiph.2012.08.004.
21. Skiada, A.; Petrikkos, G. Cutaneous Mucormycosis. *Skinmed* 2013, (3), 155–159. PMID: 23930354.
22. Sarrami, A. H.; et al. Fatal Disseminated Mucormycosis in an Immunocompotent Patient: A Case Report and Literature Review. *Int. J. Prev. Med.* 2013, *4* (12), 1468–1471.

23. Alam, A.; et al. Sinonasal Mucormycosis: Diagnosis Using Computed Tomography. *Med. J. Armed Forces India* 2003, *59* (3), 243–245. DOI: 10.1016/S0377-1237(03)80018-7
24. Suryanarayanan, T. S.;Shaanker, R. U. COVID-19 and Black Fungus: What Is Mucormycosis? [Online] 2021. https://science.thewire.in/the-sciences/covid-19-and-black-fungus-what-is-mucormycosis/ (accessed May 20, 2021)
25. Runwal, P. A Rare Black Fungus is Infecting Many of India's COVID-19 Patients—Why? [Online] 2021. https://www.nationalgeographic.com/science/article/a-rare-black-fungus-is-infecting-many-of-indias-covid-19-patientswhy (accessed May 20, 2021)
26. Carbajal, E. 6 Things to Know About 'Black Fungus' Infections Among India's COVID-19 Patients. [Online] 2021. https://www.beckershospitalreview.com/public-health/6-things-to-know-about-black-fungus-infections-among-india-s-covid-19-patients.html (accessed May 15, 2021)
27. Virtual Soil Science Learning Resources Channel. Fungi. [Online] 2012. https://biology.soilweb.ca/fungi/#:~:text=Saprophytic%20fungi%20decompose%20nonliving%20organic,fungi%20can%20also%20decomposed%20lignin (accessed May 21, 2021)
28. Bhandari, J.; Thada, P. K.; Nagalli, S. Rhinocerebral Mucormycosis. [Online] 2020. https://www.ncbi.nlm.nih.gov/books/NBK559288/#:~:text=The%20causative%20agent%20of%20the,Cunninghamella%20genera%2C%20and%20Apophysomyces%20elegans (Accessed May 12, 2021)
29. Boddy, L. Interactions with Humans and Other Animals. In *The Fungi*; Academic Press, 2016; pp 293–336. https://doi.org/10.1016/B978-0-12-382034-1.00009-8.
30. Kathryn Weber BellaOnline's Cleaning Editor. Household Dust -- What It is, Where It Comes From, and What to Do About It. [Online] 2021. http://www.pagespersonalcleaning.net/household_dust.htm#:~:text=Dust%20is%20made%20up%20of,skin%20flakes%20that%20humans%20shed (accessed May 20, 2021)
31. Bioclean. How to Know the Difference Between Dust and Mold. [Online] 2021. https://www.biocleanct.com/blog/how-to-know-the-difference-between-dust-and-mold#:~:text=If%20you%20place%20a%20few,it's%20just%20dust%20or%20dirt.&text=Mold%20doesn't%20like%20well,mold%20but%20dust%20or%20dirt (accessed May 12, 2021)
32. The Indian Express. Explained: Why it Might be a Good Idea to Keep AC Temperature Above 24 Degrees This Summer. [Online] 2021. https://indianexpress.com/article/explained/coronavirus-air-conditioner-temperature-guidelines-humidity-6378739/#:~:text=The%20guidelines%20state%3A%20%E2%80%9CSet%20room,cent%20and%2070%20per%20cent (accessed May 12, 2021)
33. The Indian Express. Quixplained: What is Mucormycosis or Black Fungus in Covid-19 Patients? [Online] 2021. https://indianexpress.com/article/explained/mucormycosis-black-fungus-covid-patients-explained-quixplained-7323887/ (accessed May 22, 2021)
34. Nivedita, R. Black Fungus: Why is it Dangerous And How to Protect Ourselves From it | Doctor Explains. [Online] 2021. https://www.india.com/health/black-fungus-why-is-it-dangerous-and-how-to-protect-ourselves-from-it-doctor-explains-4677817/ (Accessed May 21, 2021)

CHAPTER 10

Healthcare and Cybersecurity: Analysis of the UAE Healthcare System

EDYTA SKIBINSKA, KHALID ATIQ HASSAN MUBARAK, and NOOR AL ALI[3]

American University of the Emirates, Dubai, UAE

ABSTRACT

Cybersecurity is becoming an essential concern in the digital world. This is expressly evident in the healthcare sector, with the value that healthcare records hold in the wealth of personal information—the information that can put patients at risk of a data breach. Healthcare providers are considered the custodians of patient information, and they are required to be vigilant about who has access to this knowledge and to ensure medical health records' security. In accordance with these facts, the UAE has taken steps to not just protect patient information with legislation but also began its vision to provide a unified records system through the Riayati project to mitigate the risk of medical error and to make the healthcare system more efficient and effective by providing healthcare service providers with accurate, up-to-date patient's information. The chapter presents a case study of the implementation of the Riayati project in the UAE to increase healthcare quality, safety, and cost. Interviews, reading, and observations were used to collect data, allowing researchers to immerse in the context of the subject to observe their reality and gather information. The authors used this research method because the topic is understudied and it serves as a starting point for future research.

10.1 INTRODUCTION

Despite the broad implementation of electronic health records (EHRs), according to the literature (Jones, Rudin, Perry, and Shekelle, 2014; Wahls,

Global Digital Transformation and the Covid-19 Pandemic.
Rajdeep Chowdhury, K. Martin Sagayam, Neyara Radwan, Ahmad A. Elngar (Eds.)

Haugen, Cram, 2007) when it comes to improving patient safety, digital data are rarely used. Moreover, EHRs have not been designed with the goal of monitoring and measuring patient safety, limiting EHRs' ability to lower the incidence of adverse events (Jha and Classen, 2011).

Patient safety research and how to mitigate the risk of medical error and make the healthcare system more efficient and effective are still needed to discover critical safety issues and undertake extensive reviews of healthcare delivery outcomes using EHR data sources, which contain enormous volumes of consolidated, continuous clinical data.

Better tools to monitor care and recognize gaps in medical care could be part of these reviews. EHRs provide the ability to solve several recognized safety hazards using data that are currently routinely documented (Bates, Evans, Murff, Stetson, Pizziferri, Hripcsak, 2003, Szekendi, Sullivan, Bobb, et al., 2006). They can use reminders prompted by risk indicators, such as patients who are on psychiatric drugs, to identify patients at risk of falling. EHRs can help detect possible adverse events, which are generally unreported. Potential adverse events, which are often undetected, can also be recognized using EHRs. In addition, unlike retrospective data such as claims reports, EHRs enable real-time clinical data mining, allowing for the diagnosis and analysis of errors and prevention before harm is done. Clinical data also better reflect how patients were treated than administrative data (Farmer, Black, and Bonow, 2013), which are billing-optimized. For patient safety improvement efforts, EHR data may be richer—data that are more responsive and accurate than invoicing or claims data (Tang, Ralston, Arrigotti, Qureshi, and Graham 2007). Even though the time is ripe to do so, only a few healthcare systems have successfully converted this large amount of data into meaningful information for system improvement (Raghupathi and Raghupathi, 2014).

Cybersecurity is an even more important topic in healthcare, with the need to protect both a provider's network and a patient's information from disruption or theft from a cyberthreat.[4,31]

This chapter presents a case study to show real-world processes in the Riayati project execution. This chapter will give an overview of why cybersecurity is important in healthcare, as well as devices that can be vulnerable to cyberattacks, recent cyberattacks in the healthcare industry, network security methods, the UAE's response to cybersecurity as it relates to securing patients and healthcare data, and the Riayati project's significance in developing a unified health records system.

10.2 METHODOLOGY

Interviews, reading, and observations were used to collect data, allowing researchers to immerse in the context of the subject to observe their reality and gather information. The authors used this research method because the topic is understudied and it serves as a starting point for future research. The scope of this chapter will cover the importance of adequate cybersecurity practices in healthcare, highlight the value of health record data to outside forces, the UAE's response to managing patient data, legislation pertaining to the governance of patient data in the UAE, and outline the creation of IT solutions to unify and regulate health record data in the UAE.

10.3 RESEARCH QUESTIONS

How does the Riayati project relate to cybersecurity and patient care?

What is the current state of the Riayati project, and when is it expected to be live in the UAE healthcare system?

How does cybersecurity relate to healthcare and patient safety?

How would the Riayati project function as a mechanism for UAE data protection in the medical field?

10.4 OBJECTIVE AND RESEARCH PURPOSE

This chapter will provide an overview of why cybersecurity is essential to healthcare, detailing devices that can be vulnerable to cyberattacks, recent cyberattacks in the healthcare industry, and methods to safeguard networks. Furthermore, it highlights the UAE's response to cybersecurity as it pertains to securing patient and healthcare data and developing a unified health records system with the Riayati project and its significance.

10.5 ELECTRONIC HEALTHCARE RECORDS AND SECURITY CONCERNS IN HEALTHCARE

EHRs are computerized patient records that store all relevant patient information therein, such as personal, financial, and medical information; all of which is the care provider's responsibility to keep secure to prevent the theft or misuse of patient data.[5,11] As explained by Coventry and Barnley, patient care can be improved on a macrolevel through increased connectivity and ease of accessibility, thus reducing error, duplication, and unnecessary testing (Coventry and Barnley,[6] pp 2–3).

A major concern in the healthcare sector regarding security is that it tends to fall behind compared to other industries. This lagging behind is due to rising costs and difficulties with interfacing newer security measures with legacy equipment compared to how fast threats and technology advance (Kruse et al,[14] p 2). These security concerns manifest themselves in theft of patient information, interference with operations with denial of service, and ransomware, which can extend devices on patients that could put their health at serious risk. Ransomware is a kind of computer virus that functions by preventing access by a user to the network or device until the party agrees to pay the offending party to regain control of the system; this sort of disruption is a great concern in healthcare as it also puts patients and their information at risk with no guarantee that the hacking party has not copied patient information or laid additional viruses in the network.[1,3,13] The vulnerabilities in a healthcare system exist as a result of a necessity to be accessible to many users within a care provider's network, difficulties or resistance by healthcare staff to adopt new processes, misplacing of equipment with sensitive information, a general lack or limitation in the security of healthcare equipment, and improper practice of computer safety, putting the whole network at risk.[5,31]

10.6 INFORMATION STORED IN HEALTH RECORDS

A patient's medical records are a diverse wealth of information containing their personal history, demographic information, financial information, diagnostic information, treatment and medication, and any notations from care providers.[29] This information is utilized to determine the causation and resolution of complaints to the care providers and billing to the payer, such as insurance, and can be used for research purposes related to the patient or area of research. An example of how healthcare information is utilized is shown in Figure 10.1.

Patient Related	Physician Related	Support Services Related	Administrative Related
Patient History Insurance Information Contact Information	Presenting Complaint Procedure Ordering Prescribing Medication Medical Examination	Fulfilling Prescription Imagining Laboratory Testing	Processing Billing (Insurance, Out-of-Pocket, ect) Managing Patient Booking Managing Patient Discharge Managing Patient Record Data

FIGURE 10.1 An example of healthcare information organized by relevant sections.

10.7 EXAMPLES OF DEVICES AT RISK IN HEALTHCARE

Examples of common healthcare devices that can be at risk are heart regulator devices, drug pumps, and other networked equipment. Pacemakers and ICDs (implantable cardioverter defibrillators) are implantable devices that are used to regulate a patient's heart rate. If one of these devices is compromised, a patient's life could be at risk by the hacker, inducing a lethal shock to the person's heart or somehow damaging the battery. Drug administration pumps deliver intravenous medications over time. In the event that a pump has been hacked, the medication linked to the devices could be used to induce an overdose or make the patient otherwise unwell. A hospital's network and devices connected to it, if affected by a malicious virus such as malware or ransomware, could impact a care provider's ability to provide service, access information, and put a patient's information at risk.[1,16]

10.8 EXAMPLES OF CYBERSECURITY BREACHES IN HEALTHCARE

Davis reports on several breaches that occurred in the United States in 2020. Blackbaud operates cloud-based networking for a number of providers across the United States. Over a period of 3 months, the company suffered an incursion, leading to over 10 million records being breached. The company's security team was able to stop the hackers from accessing more information by locking them out, but not before names, contact information, and some medical information were unlawfully accessed, and the breach was disclosed. Luxottica, an eyecare company operating in the United States, experienced a significant breach that led to the theft of insurance data, financial data, personal information, and prescription information. These attacks illustrate the depths to which hackers go to acquire healthcare information.[9]

10.9 SAFEGUARDING HEALTHCARE NETWORKS

Network security can be improved through four strategies, as per Buntz. Maintain a basic level of security through the use of passwords and security setups. Install and maintain firewalls in the network and manage the level of interconnectivity of devices within the network. Expand the existing security system to include a layered defense system such as encryption and authentication to slow down threats and allow for time so security can isolate or protect the afflicted system. Authorization controls who has access and tracks where that access was made from; it can function as a gatekeeping mechanic to ensure only selected people, such as hospital staff, have access to the system or area.[2] Overall, a degree of versatility and sophistication is important both to strengthen the level of security in the healthcare sector and to deter potential incursions due to the effort and risk involved with penetrating those defenses.

In the event of a breach or incursion into a care provider's network and patient information at risk, there needs to be transparency, full disclosure to the public, and cooperation with the proper authorities, such as the Ministry of Health and the police. Furthermore, the healthcare sector, in general, should reevaluate the security levels and overall vulnerability of care providers in order to ensure patient safety and network security.[5]

10.10 CASE STUDY: CYBERSECURITY AND THE UAE HEALTHCARE SYSTEM

10.10.1 LEGISLATION AND PROTOCOLS

The Ministry of Health and Prevention (MoHaP) explains that the UAE's national cybersecurity strategy covers five areas to protect information and communication in the country. Those areas are improving basic cybersecurity standards in the country, improving risk management against hacking or other cyber threats, awareness campaigns related to cybersecurity, working with international bodies to better combat cyberthreats, and forming an oversight authority responsible for monitoring and addressing cyber concerns in the country.[19] This strategy aims to improve standards and educate those in the country about the importance and risks associated with cybersecurity, or lack thereof. This is crucial in healthcare, as mentioned earlier, due to the vulnerability to threats that could affect care providers and patients alike.

In Abu Dhabi, the Department of Health (DoH) has issued guidelines and regulations for the management of patient data, authorization, and access to this information. The strategy stresses the need for proper computer etiquette and that only authorized individuals should have access to patient information or a care provider's network. In addition, the strategy emphasizes that care providers are the caretakers of a patient's information, as a patient or authorized representative has the right to view their information, and that it is the care provider's responsibility to set and uphold standards in line with the principles set by MoHaP and the DoH regarding data and network security (DoH,[10] pp 21–26).

Dubai Healthcare City (DHCC) redesigned the data protection regulations of the free zone in 2013. Firstly, moving oversight to the purview of the board of directors, regulation compliance and formulation to a specific department entity, the Centre for Healthcare Planning and Quality, and the creation of the Central Governance Board, which governs patient education, legal matters related to patient data, and compliance with the rules and regulations set by the DHCC relating to patient data. Furthermore, these changes aim to improve service and care within the free zone and improve overall security regarding patient data and the ease of the transfer of said data along secure channels.[26]

In 2019, Federal Law (2) of 2019 related to information technology and communication in healthcare was enacted. This law guides the security, use, and exchange of patient information in the healthcare setting of the UAE. This law encompasses care providers, insurance providers, and support services providers, including IT. This law further regulates the transfer of patient data out of the country and sets the Ministry of Health and Prevention as the main authority that will be regulating and managing said data, as well as detailing how patient data are to be shared and addressing the punishments for mishandling patient data.[25]

10.11 RIAYATI PROJECT AND UNIFIED HEALTH RECORDS

Healthcare records are vital information sources that illustrate a patient's overall health history, with electronic records paving the way to eliminate errors and duplication in record information, which in turn will lead to better patient outcomes with a clearer picture of a patient's health history. Unified records improve patient care even more as they place healthcare providers on the same level and in several smooth areas, like care, billing, and data collection.[39]

Furthermore, unified records help with mitigating potential legal liability through added transparency of information between other care providers to reduce or remove the risk of error that may lead to malpractice occurrence.[30]

With a unified system, the effectiveness of treatments or procedures can be more easily monitored and tracked. With access to a complete source of information, better and more positive predictions of outcomes can be made. From a legal aspect, care providers are able to place themselves at less risk due to the information being under greater oversight. Errors can be further prevented by the information being up-to-date and more complete, thus improving the quality of care and patient outcomes.[30]

In addition, unified records, besides minimizing risks, also allow for streamlined and efficient care while mitigating costs in healthcare. These records also make accessing data for medical or scientific research significantly easier to collect and analyze.[24] This can be visualized in Figure 10.2.

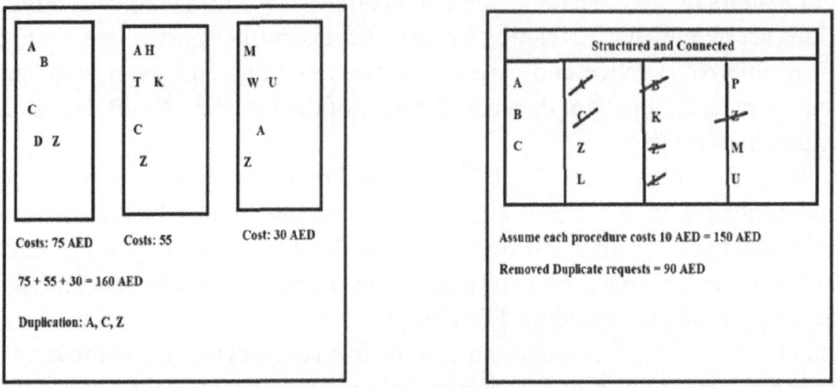

FIGURE 10.2 An example of a disjointed record (left) and a structured and connected system (right), allowing for cost control and removal of duplication.

10.12 UNIFIED HEALTHCARE RECORDS IN THE UAE

In Dubai, the Dubai Health Authority (DHA) began two initiatives, Nabidh and Salama, which will connect DHA-affiliated as well as private hospitals and clinics together under one system, allowing for better and safer care in Dubai, with care providers having access to the same information regarding the patient in question. Furthermore, the DoH has also created a similar independent system called Malaffi to connect healthcare providers in Abu Dhabi and use the system to assist in managing public health and patient diagnosis.[7,8,28] The Riayati project will connect many healthcare providers across the country and link Nabidh, Salama, and Malaffi under its scope to assure unilateral care in the healthcare sector.[18]

10.13 RIAYATI PROJECT AND CURRENT DEVELOPMENT

At Arab Health 2018, the Riayati project was announced by the Ministry of Health and Prevention to connect healthcare records nationwide as part of its strategy to improve and elevate the country's healthcare IT and infrastructure in accordance with international standards. The system will connect the existing initiatives present in Abu Dhabi and Dubai with the new system connecting the Northern Emirates. The project's main aim is to improve and promote patient care and satisfaction. The network will provide more integration and free flow of patient information among care providers, as well as improved public health monitoring and planning in healthcare.[37] In addition, the Riayati system will make access to patient data, such as previous medical interactions, medications, and insurance information, more flexible and user-friendly. The system makes efforts to improve medical decision-making and efficiency, reducing errors with duplication and inaccuracy in medical records, and reducing overall care costs.[38]

The Riayati system will utilize the HL7V2 standard for codification and internal messaging and notation within the healthcare setting for the exchange of medically and patient-relevant information.[12,32] A supporting article in June of 2020 outlines the access, utilization, and consenting process for the system, with opt out being the default, meaning each time a person interacts with a provider, their consent is required to access the data saved for emergency cases or approved research; in this case, the data are masked with personal information not relevant to the study. In addition, the article explains the process if a breach were to occur by investigating the breach, preventing the level of damage, preventing future breaches, and notifying MoHaP.[33]

In March 2021, the MoHaP held a webinar directed toward healthcare providers, where the ministry reiterated the purpose and benefits of the system and how it will actualize the UAE's healthcare vision for 2021 to provide protected and accurate patient data that are accessible to the medical community for the purpose of managing and treating patients. The webinar outlines the benefits the Riayati system will have over conventional record-keeping and the advantages to the quality of patient care the system will provide. In addition, the webinar outlined how care providers could utilize the system and the current plan to facilitate the system changeover.[20] In early September of 2021, the ministry released an announcement regarding the technical aspects of the system and how the information would be accessible and managed. The system would run on SNOMED CT, which is specifically designed to cater to medical data used across 60 countries; this language is praised for being multifunctional in medical data management as well as being comprehensive to be utilized in a universal setting, meaning that the information is easily translatable to systems that operate with it even internationally, where language can be a barrier.[21] Dr. Abdulaziz Al-Zarouni, acting assistant undersecretary for the Support Services Sector, indicated that the change is a revolutionary step forward to a universal records system as it assures all parties involved with patient care are equally informed, as it would be breaking down language barriers to build a more common and equal footing for the transfer of information and ultimately improve patient care. At the end of September, the ministry held another series of announcements. The most major of which relates to the expected launch of the service to be operational by January 2022. During the seminars, the ministry detailed how the management of insurance claims and payment acquisition would be handled by Pure CS through a dedicated claims channel. This partnership would streamline the patient experience, management of insurance claims, and resource utilization to prevent misuse or the wasting of valuable resources and time.[22]

As of March 1, 2022, the Riayati website portal[35] reports have connected 7.6 million patient files, 244 healthcare facilities, and over 30,000 clinicians. From the connected participants being a mix of testing facilities, hospitals, and clinics, they only represent a fraction of the total number of providers and support centers. As per government statistics, the most recent statistics are from 2017; there are 5111 clinics and hospitals across the country, and the current participating facilities only constitute 4.7%. The Riayati website indicates that the participating facilities have connected records involving patient medical information, including patient vital signs, vulnerabilities, medical procedures, and various laboratory results, saved for microbiology results.[34,36]

In an interview during Arab Health 2022, the MoHaP representative explained that the system is still underway and that there is some delay in connecting Abu Dhabi and Dubai's record systems together. Connecting the different existing regional record systems to the Riayati project's mainframe has been a significant issue for the project. The current goal of MoHaP is to have the system go live within the year or sometime in 2023.[23]

10.14 OVERSIGHT AND RECORD ACCESS

With reference to Federal Law (2) of 2019, the governing authority for healthcare data storage and management is the Ministry of Health and Prevention, and the law states how care providers will utilize and access patient data, along with how misuse of the said information would be punished. In addition, the law regulated the duration that information would be kept for historical reference, the extension of the liability to third-party agents to healthcare, and how patient information would be used for medical or scientific research.[17]

10.15 COMPARISON

TABLE 10.1 Highlighted Benefits of an Integrated

Health System Based on a Study by Kurpas et al.[15] Highlights of an integrated health system
Greater effectiveness and efficiency
Reduced error, duplication, and costs
Improved billing and insurance integration

In a study across central and eastern Europe, it was shown across the nine participating countries—Poland, Belarus, Romania, Ukraine, Russia, Slovakia, Hungary, Lithuania, and the Czech Republic—that integration improves healthcare outcomes such as patient health and staff efficiency by improving information access and reducing error. Among those countries, Hungary and Romania have the highest level of integration across multiple areas, with the other countries having variable levels of integration primarily in support services and insurance integration rather than in primary care. One of the main issues holding back wider-scale integration in primary care in those regions is IT solutions and proper training in utilizing those sorts of systems. The study highlighted the importance of greater integration and funding to better manage aging populations and emerging conditions within the population.[15]

10.16 CONCLUSION AND DISCUSSION

To summarize, cybersecurity is a very crucial part of a healthcare system, as without an adequate level of security, patient safety, health, and financial security can all be at risk. To combat unwanted theft of data or disruption of care by hackers, good system etiquette should be followed, such as making sure the system's security is secure, ensuring that any device that could contain patient data is accounted for, and regulating the authorization and access to the system by individuals and devices. With regards to the security of the network or device, it should be multifaceted in order to provide the best protection and adequate time for the IT department to handle a protentional incursion and to be open and transparent to affected parties regulators, and the public in the event of a breach. With regards to the UAE, several regulatory bodies had tried to form a standard of care and procedure surrounding the security of health data before a law was enacted to provide clearer and defined standards to uphold the security of patient data and public as well as private safety. The Riayati project is the UAE's solution to developing a unified system to provide secure and accurate access to patient information to healthcare providers. This system aims to improve the quality of patient care by providing accurate, up-to-date patient data, mitigating errors, and securing financial exchanges within the healthcare environment. The system will utilize a universal coding and language system that will allow for easily translatable information that will not be hindered by a language barrier. It is currently in its late stages of development, with an expected launch by the first quarter of 2022. As the system is a government-supported project managed and implemented by the Ministry of Health and Prevention, it will regulate who has access to patient records and will be able to monitor such access as it complies with the legislation on accessing and utilizing patient data.

Considering the growing importance of cybersecurity for the proper functioning of the healthcare sector, it becomes necessary to comprehensively study the issues discussed in the literature on the borders of medicine and cybersecurity. We believe that the cooperation of both of these sectors can bring tangible benefits to all interested parties. Due to insufficient data on the cybersecurity market in healthcare, it will be necessary to conduct extensive research on the subject in the future. It would be important to assess the awareness of medical and administrative staff, as well as their daily behavior in this regard.

10.17 IMPORTANCE OF A UNIFIED SYSTEM

Without connectivity, a patient's records can be fragmented, incomplete, or erroneously duplicated among different care providers. This can lead to a patient repeating their history on many visits to different specialists or care providers, thus leading to delays in treatment or having to repeat examinations that may have been done already; otherwise, the patient would have to bring copies of their records from other institutions. Adopting a unified system lessens such burdens on the patient and reduces the risk of medical error and excess spending.[27] Unified systems such as Malaffi, Nabidh, or Riayati will be crucial to the continuum of care as they streamline caregiving, reduce unnecessary costs and testing, and allow for a more efficient flow of information overall, leading to better patient outcomes and high-quality delivery of patient care.

KEYWORDS

- **cybersecurity**
- **healthcare**
- **sensitive data**
- **electronic health records**
- **cyberattack**

REFERENCES

1. Alpine Security. Most Dangerous Hacked Medical Devices [Online], ND. https://alpinesecurity.com/blog/most-dangerous-hacked-medical-devices/
2. Buntz, B. Tips to Keep Your Medical Device from Getting Hacked [Online], 2013. https://www.mddionline.com/software/tips-keep-your-medical-device-getting-hacked
3. Centre of Internet Security (CIS). Cyber Attacks: In the Healthcare Sector [Online], ND. https://www.cisecurity.org/blog/cyber-attacks-in-the-healthcare-sector/
4. Cybersecurity and Infrastructure Security Agency (CISA). What is Cybersecurity? [Online], 2019. https://us-cert.cisa.gov/ncas/tips/ST04-001
5. Consoltech. Security Threats in HealthCare Systems [Online], 2019. https://consoltech.com/blog/security-threats-healthcare-systems/
6. Coventry, L.; Branley, D. Cybersecurity in Healthcare: A Narrative Review of Trends, Threats and Ways Forward. *Maturitas* 2018, *113,* 48–52.
7. DHA. Dubai Health Authority Launches Information Security Project [Online], 2017. https://www.dha.gov.ae/en/DHANews/pages/dhanews762953541-30-07-2017.aspx
8. DHA. DHA Introduces NABIDH, Welcoming the Future of Healthcare Excellence in Dubai [Online], 2020. https://www.dha.gov.ae/en/DHANews/Pages/DHANews1904903097-30-11-2020.aspx
9. Davis, J. Update: the 10 Biggest Healthcare Data Breaches of 2020 [Online], ND. https://healthitsecurity.com/news/the-10-biggest-healthcare-data-breaches-of-2020
10. Department of Health (DoH). Policy on the Abu Dhabi Health Information Exchange (HIE) [Online], 2020. https://www.doh.gov.ae/en/resources/policies
11. HealthIT.Gov. What is an Electronic Health Record (EHR)? [Online], 2019. https://www.healthit.gov/faq/what-electronic-health-record-ehr#:~:text=An%20electronic%20health%20record%20(EHR)%20is%20a%20digital%20version,of%20a%20patient%27s%20paper%20chart.&text=EHRs%20are%20a%20vital%20part,and%20laboratory%20and%20test%20results
12. HL International. HL7 Version 2 Product Suite [Online], ND. https://www.hl7.org/implement/standards/product_brief.cfm?product_id=185
13. HIMSS. Cybersecurity in Healthcare [Online], ND. https://www.himss.org/resources/cybersecurity-healthcare

14. Kruse, C.; Frederick, B.; Jacobson, T.; Monticone, D. Cybersecurity in Healthcare: A Systematic Review of Modern Threats and Trends. *Technol. Health Care* **2017,** *25,* 1–10.
15. Kurpas, D.; Stefanicka-Wojtas, D.; Shpakou, A.; Halata, D.; Mohos, A.; Skarbaliene, A.; Dumitra, G.; Klimatckaia, L.; Bendova, J.; Tkachenko, V. The Advantages and Disadvantages of Integrated Care Implementation in Central and Eastern Europe—Perspective from 9 CEE Countries. *Int. J. Integr. Care* **2021,** *21* (4), 14.
16. McKinley, J. 5 Medical Devices You didn't Know Could be Hacked [Online], ND. https://www.thehealthy.com/healthcare/hacked-medical-devices/
17. Ministry of Health and Prevention (MoHaP). Federal Law No. (2) of 2019 Concerning the Use of Information and Communication Technology (ICT) in Health Fields [Online], 2019. https://www.mohap.gov.ae/en/Aboutus/Pages/PublicHealthPolicies.aspx
18. Ministry of Health and Prevention (MoHaP). Ministry of Health and Prevention to Connect "Riayati" Platform with Other Medical Records Including 'Nabidh' & 'Malaffi' [Online], 2020. https://www.mohap.gov.ae/en/MediaCenter/News/Pages/2725.aspx
19. Ministry of Health and Prevention (MoHaP). Digital Security [Online], ND. https://www.mohap.gov.ae/en/Pages/NCSS.aspx
20. Ministry of Health and Prevention (MoHaP) A. Ministry of Health Introduces Riayati Platform to Healthcare Service Providers [Online], 2021. https://www.mohap.gov.ae/en/MediaCenter/News/Pages/2843.aspx
21. Ministry of Health and Prevention (MoHaP) B. MoHAP Adopts Advanced Technology Solutions to Collect and Encode Healthcare Data within 'Riayati' Platform [Online], 2021. https://www.mohap.gov.ae/en/MediaCenter/News/Pages/3081.aspx
22. Ministry of Health and Prevention (MoHaP) C. Ministry of Health and Prevention Introduces eClaims Office Service to Insurance Companies, Healthcare Service Providers [Online], 2021. https://www.mohap.gov.ae/en/MediaCenter/News/Pages/3108.aspx
23. Ministry of Health and Prevention (MoHaP). Riayati Programme Regarding Unified Healthcare Records [Interview], 2022.
24. Norden, J. G.; Wang, J. X.; Desai, S. A.; Cheung, L. Utilizing a Novel Unified Healthcare Model to Compare Practice Patterns Between Telemedicine and In-Person Visits. *Digital Health* **2020,** *6,* 2055207620958528.
25. PWC Middle East. Healthcare Data Protection in the UAE: A New Federal Law [Online], ND. https://www.pwc.com/m1/en/publications/healthcare-data-protection-in-the-uae.html
26. O'Connell, N. Protection of Patient Health Information in Dubai Healthcare City [Online], 2014. https://www.tamimi.com/law-update-articles/protection-of-patient-health-information-in-dubai-healthcare-city/
27. Otto, L.; Harst, L.; Schlieter, H.; Wollschlaeger, B.; Richter, P.; Timpel, P. In *Towards a Unified Understanding of eHealth and Related Terms-Proposal of a Consolidated Terminological Basis*, HEALTHINF, 2018; pp 533–539). https://care4saxony.de/wp-content/uploads/2018/05/2018_HEALTHINF_Otto.pdf
28. Salman, N.; Alfaham, T. Abu Dhabi, Dubai Partner to Develop Global Healthcare Model [Online], 2020. https://www.wam.ae/en/details/1395302823117
29. Souza, M. D. F. M. D.; Malta, D. C.; França, E. B.; Barreto, M. L. Changes in Health and Disease in Brazil and Its States in the 30 Years Since the Unified Healthcare System (SUS) was Created. *Cienc. Saude Colet.* **2018,** *23,* 1737–1750.

30. Subasi, A.; Radhwan, M.; Kurdi, R.; Khateeb, K. In *IoT Based Mobile Healthcare System for Human Activity Recognition*, 2018 15th Learning and Technology Conference (L&T), IEEE. 2018; pp 29–34. https://www.researchgate.net/profile/Louis-Hickman/publication/325495109_Developing_intercultural_competencies_through_virtual_reality_Internet_of_Things_applications_in_education_and_learning/links/60404651299bf1e07854448e/Developing-intercultural-competencies-through-virtual-reality-Internet-of-Things-applications-in-education-and-learning.pdf#page=36
31. SwivelSecure. 9 Reasons Why Healthcare is the Biggest Target for Cyberattacks [Online], ND. https://swivelsecure.com/solutions/healthcare/healthcare-is-the-biggest-target-for-cyberattacks/
32. Riayati Ministry of Health and Prevention. Riayati Program Integration, Terminology and Coding Standards [Online], 2020. https://riayati.mohap.gov.ae/resources/Riayati_Integration_Terminology_and_Coding_Standards_v2.0.pdf
33. Riayati Ministry of Health and Prevention. Riayati Confidentiality and Data Privacy Policy [Online], 2020. https://riayati.mohap.gov.ae/resources/Riayati_Confidentiality_and_data_privacy_policy_V0.7.pdf
34. Riayati Ministry of Health and Prevention. Health Information Exchange [Online], 2021. https://riayati.mohap.gov.ae/HIE.html
35. Riayati Ministry of Health and Prevention. Connected Participants [Online], 2022. https://riayati.mohap.gov.ae/connected-participants.html
36. UAE Government. Healthcare Providers [Online], 2021. https://u.ae/en/information-and-services/health-and-fitness/healthcare-providers
37. Zawya. Ministry of Health to Connect "Riayati" Platform with Other Medical Records Including "Nabidh" & "Malaffi," [Online], 2020. https://www.zawya.com/mena/en/press-releases/story/Ministry_of_Health_to_connect_Riayati_platform_with_other_medical_records_including_Nabidh__Malaffi-ZAWYA20201228100110/
38. Zawya. Riayati Platform Introduced to Healthcare Service Providers [Online], 2021. https://www.zawya.com/mena/en/life/story/Riayati_platform_introduced_to_healthcare_service_providers-SNG_206009924/
39. Zhang, Q.; Bai, C.; Yang, L. T.; Chen, Z.; Li, P.; Yu, H. A Unified Smart Chinese Medicine Framework for Healthcare and Medical Services. *IEEE/ACM Transactions on Computational Biology and Bioinformatics* [Online], 2019. https://ieeexplore.ieee.org/abstract/document/8704881/

CHAPTER 11

Social Media and Technology Instruction Design in Teaching Amid COVID-19

MUTHMAINNAH[1], AHMAD AL YAKIN[1], ANDI ASRIFAN[2], and AHMED A. ELNGAR[3]

[1]*Universitas Al Asyariah Mandar, West Sulawesi, Indonesia*

[2]*Universitas Muhammadiyah Sidenreng Rappang, South Sulawesi, Indonesia*

[3]*Faculty of Computers and Artificial Intelligence, Beni Suef University, Beni Suef City, Egypt*

ABSTRACT

The standard lecture and learning methods do not significantly improve students' cognitive and communicative abilities in higher education. Modern ICT tools are available to improve the teaching–learning process and the efficacy of the education system. Examination of the literature reveals that incorporating current ICT tools into teaching and learning processes significantly improves teaching and learning processes. The COVID-19 epidemic has changed people's lives in all fields, including higher education universities. To comply with the health protocol, the learning process cannot be done face-to-face; instead, it is done primarily online. This chapter aims to examine contemporary information and communication technology tools and their use in the teaching and learning process. Therefore, this chapter seeks to examine how the learning process is carried out in online learning-based education study programs. The research sample consisted of 43 respondents who were randomly selected from 12 study programs. The data collection instrument used in this study was an online questionnaire, and the results were analyzed quantitatively and descriptively. The findings

Global Digital Transformation and the Covid-19 Pandemic.
Rajdeep Chowdhury, K. Martin Sagayam, Neyara Radwan, Ahmad A. Elngar (Eds.)

show that instructors in the study program often use online communication tools, namely WhaYoZo, which is the most frequently used, followed by media and technology learning tools. The results of this study suggest that universities should support and encourage SOCMED-WhaYoZo to improve technical skills, performance, expression of ideas, arguments, and student achievement for sustainable development.

11.1 INTRODUCTION

Using modern information and communication technology (ICT) in education has made it possible to conduct distance learning. An online course must provide the same knowledge and skills as a traditional classroom setting. Students can learn at their own pace, from any location in the world, at any time of day or night, and at any level of intensity they choose.[2,3,30,42]

According to Hafeez et al.[22] and Top et al.,[66] it is essential for a country's growth to have a well-functioning educational system. Because of the widespread use of ICT in education, the traditional methods of instruction have undergone significant modifications. Teachers and students both benefit from the enhanced efficiency that ICT has brought to the teaching–learning process. Teachers should also participate actively in the process of developing ICT training material.

Currently, the investigations carried out by Nikolić et al.[46] suggest that the latest ICT tools can improve the teaching and learning process efficiently. Curriculum development and instructional design have evolved as a result of the widespread use of ICT in teaching and learning. The curriculum is now being developed by educators and administrators to meet the demands of the ICT era.

Many educators and academics have cited the advantages of incorporating ICT into the teaching and learning process.[20,41,45,49,58] Samoylenko et al.[56] reported that they would be eager to take the course that focused on ICT. When students urge their lecturers to make their materials more understandable, interesting, and enjoyable, using ICT in teaching is highly desired by the students. ICT is also utilized in off-class time to push the learning content actively and maximize the success of mobile learning. It was possible to create a continuous blended spatial system with intelligent, tailored, diverse mobile learning and management by expanding standard classroom learning to an "off-class autonomous inquiry learning, class information absorption, after-school knowledge push" model. They also claim that using ICT to create educational materials is a great way to get students engaged emotionally, behaviorally, cognitively, socially, and culturally.

The way students are taught and educated has been profoundly altered by the rapid advancements in ICT and the internet. The incorporation of ICT into the teaching–learning process improves students' comprehension of courses and subjects. With a wide range of pedagogies and materials to choose from, the teaching–learning process becomes easier and more efficient.[31] Teaching and learning in the 21st century are more impacted than ever before by advances in ICT. Another benefit of using ICTs is an improved educational system.

Zourmpakis et al.[76] found that the adoption of contemporary ICT in education facilitated the development of a new style of education known as remote learning. This mode of education must provide the same amount of information and abilities as traditional education. The fundamental feature of this kind of education is that it enables students to learn at any time, from any location on the globe, and at their own pace.

Additionally, emerging digital technologies have created a new paradigm for higher education and professional training. Picciano,[53] the first to begin technology and digitalization has altered how young students see and process information, their patterns of engagement during the educational process, and their day-to-day communication. Second, the current educational system places a premium on lifelong learning. Through involvement and interaction in new learning settings, online learning facilities provide numerous chances for students to acquire new knowledge and enhance their abilities. The adoption of the "open education" idea has altered the way learners acquire, share, and consolidate information since they now have almost unrestricted access to high-quality education and learning resources. Thus, in a digital world, the mere transmission of information is less important than the development of students' creative potential, professional abilities, and competencies via the use of technology and face-to-face interactions. As a result, facilitative contact is often considered the predominant tendency in higher education.

These characteristics contribute to e-appeal and learning, as does the widespread usage of ICT and other technologies for remote education. As a result, a variety of e-learning technologies, blended learning models, and approaches are being formally integrated into the Indonesian educational system at all levels, starting with the renewed curriculum of Merdeka Belajar Kampus Merdeka (MBKM). In light of Indonesian educational changes, contemporary e-learning is one method for increasing student motivation, engagement, educational quality, and the efficiency of ICT as tools in teaching amid COVID-19. The use of ICT and other e-learning technologies drives the development of new flexible learning environments, cognitive behavior, metacognition skills, and 21st-century skills that enable students and educators to connect productively, which will be investigated in this chapter.

11.2 LITERATURE REVIEW

11.2.1 DIGITAL TRANSFORMATION PITFALL AND POSSIBILITY

Online learning and teaching face a variety of challenges related to the learner, the instructor, and material problems. Organizations have difficulties attracting learners and integrating them into the learning process. It is difficult for instructors to go from offline to online mode without affecting their instructional method or time management. Along with addressing the curriculum, it's difficult to create resources that incorporate students.[28] The quality of online learning curricula is a significant issue. The state's instructional methodologies make no explicit reference to online learning programs. There is a shortage of quality principles for online resource development, delivery of online learning content, and quality monitoring.

To ensure that all students have access to high-quality online education, this problem must be addressed as soon as possible. Not only should experts who provide online education during disasters receive the most attention, but also those who create and improve the quality of computer-generated courses offered during such situations.[1] Online education can be both time-consuming and expensive. Investments in technology, human capital development, and online learning materials are required to make this a reality. As a result, there is a pressing need to develop an efficient and effective online education system.

It is imperative to ensure digital equity at this critical juncture. Access to all devices, Wi-Fi, and the internet isn't available to everyone. It can be extremely inconvenient for many students to not have access to the proper digital media, internet connections, or Effie Wi-Fi connectivity. Administrations must work to ensure that all students and faculty members have access to the resources they need. If they don't have access to a laptop, students should make sure that all of their educational programs are mobile-friendly. Consequently, it is necessary to take action in order to reduce the digital gap. Students and teachers in many educational settings lack formal education in online learning. The vast majority of these issues can be dealt with and resolved in a satisfactory manner through the use of standard instructional methods, taking advantage of the current situation, which has been made worse by the coronavirus outbreak. Despite the challenges we face, we have a lot of potential to make a difference.

The teaching–learning process can be aided by a wide range of instruments. It's up to teachers to decide on the best tool for getting their message across to their students. Step-by-step instructions for instructors and students alike should be provided by educational institutions to reduce digital illiteracy by addressing major course themes via online learning resources. Video, audio, and lesson options are all available to instructors. The ability for professors to receive immediate feedback on their lectures, whether through video lectures, computer-generated meetings, or other means, is an asset.

11.2.2 HOW TO ENGAGE IGEN WITH SOCIAL MEDIA

Safko and Brake (2010) defined social media as "the mediums through which we communicate." Additionally, some academics have provided more thorough and sophisticated definitions. Westerman et al. (2014) define social media as "a broad category of channels and apps that emphasize cooperation and teamwork in the creation of information" (2014: 172). Social media, according to Kaplan and Haenlein (2010), are internet-based apps built on the Web 2.0 concept that enable and facilitate the creation and sharing of user-generated content (UGC). "Communicating, sharing, collaborating, publishing, managing, and engaging" are Mao's expanded definitions of these concepts (2014: 213). To the exclusion of older forms of media and Web 1.0 technology, social media platforms allow users to work together on the creation and editing of works (Leadbeater 2009; Shirky 2008), communicate with one another, and build communities around common interests (Boyd, 2008b; Choudhury et al., 2010; Mayfield, 2008). In this way, social media users can reach a wide audience regardless of their position in the social media hierarchy (Etter, Ravasi, and Colleoni, 2019).

Boyd (2008b: 92) defined social media as "nearly any online communication tool" that includes "messaging, blogging, microblogging, forums, email, virtual worlds, texting, or SNS" (Boyd, 2008b). When it comes to creating networked places for interaction, social media include technologies that can be used to facilitate one-to-one, one-to-many, and many-to-many communication. The social networks that people build in real life can be mirrored virtually on SNSs. Web-based services that allow individuals to create a public or semipublic profile within a constrained system, communicate with other users, and examine and traverse their own and others' relationships within the system" were defined by Boyd and Ellison (2008:.211).

Classification schemes for social media technologies have been proposed by a large number of researchers. Their nature, role, and the type of information they transmit can all be used to define them (Bozarth, 2010; Berthon et al., 2012; Kaplan and Haenlein, 2010; Mao, 2014; Xiang and Gretzel, 2009). When it comes to social media platforms, Bozarth (2010) broke them down into five distinct categories based on the goals and functions of the content they provide. These categories include: (1) communication (like blogs); (2) cooperation (like wikis); (3) communities (like Facebook, WhatsApp, and Messenger); (4) reviews (like those found on Amazon); and (5) multimedia (e.g., YouTube). According to Safko's (2012) categorization, all social media tools will be categorized by function throughout the book. The categories include: (1) social networking, (2) wikis, (3) forums, (4) blogs, (5) picture sharing, (6) audio, (7) video, (8) microblogging, (9) live casting, (10) virtual worlds, (11) gaming, (12) RSS and aggregators, (13) search, (14) mobile, and (15) interpersonal.

Their role as change agents in their schools is also to promote the adoption of new technologies. Educators use technology in two ways. As Tubin[67] points out, one of these is the ability to achieve traditional goals in the same classrooms without significantly altering the activities of the teachers and students. This method involves using technology in the classroom to remove barriers, connect students to real-world activities, and assist them in becoming more independent learners. Teachers' views on education and learning are likely to influence how they use technology. Teaching methods and classroom conduct are greatly affected by teachers' beliefs and values.

Education technology is more likely to be adopted when teachers have access to ICT. There are many ways to incorporate technology into the classroom, and these include articulating how they make academic decisions about lesson planning, implementation systems, and evaluation strategies. According to Munoz-Miralles et al.,[44] advances in ICT have led to new forms of communication, function, and education. It is advantageous to universities because it allows them to provide comprehensive and efficient training to students who will soon be technology professionals.

The use of ICT as a learning engine has been argued by experts such as Ferro Soto,[18] Orellana,[50] and [16]. Because each student learns in a unique way, effective study methods are essential to academic success. With their constant exposure to technology, students find ways to incorporate it into the education they receive. As a result, researchers like Yazon et al.[73] and others in the educational community are becoming increasingly interested in how students learn when they use ICT. It's exciting to see how ICT can be used in university education and how much it enhances the learning experience. To teach and conduct research, professors need to be trained in the use of ICT. For academic and professional success, students' views on ICT are crucial.

However, a positive evaluation of ICT components does not guarantee that they will be used in the teaching process.[4] As a starting point, Cabero Almenara J et al.[9] can be a good place to start. Even in the realm of education, ICTs are becoming more and more prevalent. As a result of having a point of contact, students' training has evolved. We can now engage in challenging formative activities that are not possible with traditional technology approaches, thanks to ICT, as Cabero points out.

The use of distance learning and online classes is on the rise. Brown et al.[8] argue that it has quickly become the preferred or "modern standard" method of research in the world. What happens to first-year distance students after they enroll in postsecondary institutions, what drives them, and how they adapt to formal distance learning are surprisingly little known. It was also found that academics' use of social media is still extremely restricted and regulated, and academics are hesitant to incorporate these technologies into their activities for a variety of reasons, including cultural resistance, pedagogical concerns, and institutional constraints. But the ways in which academics use and view social media vary widely depending on the scientific field in which they work.

Educators were compelled to develop, adapt, and implement digital educational techniques quickly in order to meet the changing needs of their students as a result of this paradigm shift. As technology for e-learning and m-learning (mobile) became more widely known, many teachers faced the challenge of learning how to use them effectively. It is possible to study from any location and at any time, as well as to think quickly and respond more quickly in an online education environment.[6]

Because of the online environment and the new pedagogical approaches, students are more engaged and have a more customized learning experience. Digitized educational resources allow students to design their own learning paths. Self-regulation and self-monitoring of academic progress are possibilities for these students. Blended learning is a popular teaching method in today's educational system. Blended learning is an essential tool for 21st century learners. Blended learning, according to the UNESCO guide, gives students both flexibility and support by combining online and face-to-face approaches. Dziuban et al.[15] and Picciano (2018) define "blended learning" as a combination of ICT, e-learning tools, and distance learning educational technology with traditional face-to-face contact (2021). Any method of delivery that incorporates technology is also commonly referred to as technology-enhanced delivery.

Blended learning, according to Hrastinski,[25] includes "...basically all forms of education that incorporate some part of face-to-face learning, and online learning is sometimes referred to as mixed learning." Some academics consider blended learning to be a new educational paradigm that combines classroom instruction with online learning. Because of this, it is still difficult to come to a clear understanding of what is meant. "Blended learning" is frequently used to describe a wide range of blends, including instructional methods, pedagogical approaches, and technologies that don't necessarily fit the accepted definition of blended learning.

It is common for blended learners to work in groups and independently in the classroom as well. Using ICT and e-learning tools in combination with flexible distance and online learning, they are able to plan their own work and interact with lecturers and fellow students in person. The trend toward networking and decentralization of learning is reflected in this style. When it comes to integrating e-learning tools and ICT into the traditional teaching methods of high school, we refer to this as blended learning.

Electronic and rechargeable devices, such as cell phones, computers, tablets, and a variety of other gadgets, disseminate social media, a form of media that allows people to share their thoughts with one another quickly and easily. It is widely used to spread information and raise public awareness all over the world. Administrators, instructors, and students may have been unable to access an e-learning platform as a result of the pandemic. The internet has undergone a dramatic shift in the last decade.[29] Many college students use social media sites like Facebook, YouTube, Instagram, WhatsApp, and Twitter for long periods of time each day. Various social media platforms exist, including these. To this day, YouTube remains Indonesia's most widely used social media platform. Almost everything we do involves the use of language as a means of exchanging ideas and information with one another. People use it to communicate with friends, family, colleagues, teachers, and students, among others Khan et al., (2016). Digital tools, including mobile devices, gaming, and social media, have become increasingly popular in the field of second and foreign language acquisition in recent years. [71]

Here are some social media platforms that instructors may utilize:

1) Facebook

Facebook users create their own personal webpages, dubbed "Profiles," using basic menus. By altering their Facebook account's privacy settings, users can restrict access to their profiles. Generally, "Friend" status grants unfettered access to a person's profile, including the ability to read and comment on all that person's galleries. To connect, users may send "Friend Requests" to other Facebook users and accept them, allowing them to view one another's profiles and comment on one another's "Walls," which are interactive blogs about their activities with feedback from others. These walls are frequently adorned with links, images, animations, and interactive items or presents. Additionally, events and parties may have their own website.[65]

Facebook is one of the most popular social networking websites. Mark Zuckerberg established it with the goal of uniting individuals from all backgrounds and encouraging involvement. Individuals use Facebook to communicate with one another. They communicate efficiently with their family, friends, and universities. "Individuals from collectivistic civilizations rely on social networking sites to build intimate relationships with a restricted number of connections rather than initiating new encounters with people." The fact that the majority of Facebook users are female is an intriguing feature of the phenomenon.

2) Weblogs

A blog is a self-published collection of text and photos that is available over the internet. A "blog," sometimes known as an "internet log, journal, or chronicle," is a web-based journal, newspaper, or chronicle maintained by an individual or group that features ongoing reporting that frequently contains diary-style comments and criticism of online publications and related news. Blogs may be a fantastic medium for students and teachers to communicate material during a course. The instructor might establish a blog to document the development of the course, session by session. Learners and others would be able to access PowerPoint presentations, lecture notes, transcripts of lectures or class sessions, and other resources (provided this is freely available on the web; Sharifzadeh et al.).[60] Blogs differ widely in their emphasis, complexity of the topic matter, and notoriety. Daily announcements, comment exchanges, and short blogs all fall within the category of the blogging tradition. Learners will communicate in a new language through blogs, an entertaining venue.

3) YouTube

As we reach the second decade of the 21st century, video is becoming more prevalent in many sectors of society, including higher education. YouTube is a massive repository of educational videos and one of the most visited websites on the internet. On the web, there are over 10 million educational films, many of which have been uploaded by peers.

A positive relationship between YouTube videos and e-learning among disabled individuals was also found by Tahat et al.[62] Disabled people also benefit from the video's high quality, user-friendliness, and educational content. Consequently, the findings strongly supported the use of technology in the global healthcare crisis. Other aspects of YouTube videos that may influence people with disabilities to use social media during an emergency like COVID-19 should be examined, according to the researchers.

4) Twitter

Twitter, the most popular free microblogging platform, is ideal for exchanging brief bits of information, expertise, and ideas.[26] The teacher will design their account in such a way that only the courses that require access to critical tweets, such as particular lectures, will have access to them. Certain professors encourage students to tweet during class sessions in order to engage and excite them about the session's discussions and subjects.

5) WhatsApp

WhatsApp is a free messaging program that is compatible with a variety of devices, including iPhones and Android phones. Undergraduate students usually use it to communicate visual communications such as images, videos, and audios, as well as basic text messages. Due to the fact that WhatsApp requires internet connectivity, a vast amount of knowledge can be gained in real time, and information transmission via technology is rapid and simple.[68]

6) Instagram

Since its inception in 2009, the cross-platform centralized messaging and voice-over-internet protocol (VoIP) service WhatsApp has offered free voice calls and text messages to millions of users around the world. It's a mobile app that can also be used on a desktop PC running Windows 8 or later or a Mac. To use WhatsApp, you'll need a smartphone or tablet running Android OS 4.0.3 or higher, or an iPhone, as well as a SIM card, an internet connection, and a phone number. User accounts are created with the phone number as the username.

WhatsApp is a messaging app that utilizes a phone's internet connection to allow users to send and receive free and limitless messages. Messages can be sent to a large number of recipients at once, thanks to this feature. Up to eight people can have secure video and audio conversations over local and international networks using the internet connection on their phone through WhatsApp. End-to-end encrypted group chats (up to 256 participants) with the capability of sharing multimedia—text and video messages, images, videos, and documents up to 100 MB—via a mobile device or desktop computer are now possible with WhatsApp Group chat. More than 2 billion people in 180 countries use WhatsApp, which is available in 60 different languages. Every day, more than 100 billion WhatsApp messages are exchanged, while more than 2 billion min are spent in video and audio chats on the messaging service.[36]

According to R. E. A. Ghobrini[21], Instagram is a photo and video sharing social networking site that was launched in October 2010, which allows users to upload and share their work. Emoji reactions and written comments allow users to have direct conversations with submitted and published content they've seen and liked by other users. With Instagram's numerous picture filters, users can quickly and easily alter the look and feel of their photos. To appeal to millennials, who have grown accustomed to a world of screens, pictures, and the like, Instagram has a unique digital design. Additionally, Instagram has a slew of other features that make it a complete social networking app, such as the ability to create profiles, add captions and tags, and access and follow the accounts of others. Users can also receive notifications, leave comments on posts, and browse the main feed, which displays the most recent content from other users' profiles.

7) LinkedIn

Socrates López-Carril and others[35] describe that it is ideal to begin with an overview of the platform's history and evolution, followed by current user stats, an explanation of the platform's primary functions, an emphasis on the platform's similarities and differences with other social media platforms, and selectively showing real examples of usage by various stakeholders in the sports industry (e.g., clubs and federations; sports brands and athletes).

8) Pinterest

According to Karimi,[27] Pinterest is a popular social networking platform with millions of members and postings that are viewed by billions of people.

There is a good chance that educators and the materials they use to educate the next generation have been buried. For instructors who utilize online social media for professional development, it is unclear how they are related to each other. As a result, understanding and facilitating teacher connections can have several benefits, such as speedier and more efficient distribution of resources. We take two significant steps in this research to better understand the connections between teachers and students. Initially, we gathered 540 instructors from across five states and 48 districts on Pinterest to create a huge dataset. Our research also includes an in-depth examination of the data to see how educators are interacting with one another on Pinterest.

As part of our research, we look into the possibility of social connections between teachers by examining a number of teacher-related traits that may be connected to the relationship a teacher forms with another one. There are a significant number of teachers on Pinterest who are only connected to other educators in their own school district or state. As a result, we compare and contrast the instructors' virtual (Pinterest) and physical (Facebook) networks and draw parallels between them.

9) TikTok

Zhu et al.[75] describe the video-sharing app through TikTok, which has clips ranging from 15 s to 5 min long. TikTok was the seventh most downloaded app in the world from 2010 to 2019 and the second most downloaded app in the world in 2019, according to data from the international market. TikTok has 40 million users in China's social media industry, making it one of the most prominent apps in 2019. On TikTok, users may make and share microvideos ranging from a few seconds to a few minutes with the TikTok community at large. It was launched in 2017 and is currently accessible in over 150 countries, making it the fastest-growing social networking app on the globe. In 2018, it topped the "Most Downloaded" ranking in the United States. One billion downloads and 500 million active users are claimed for TikTok. This social media network features brief microvideos with easy-to-use editing and music-inclusion features, unlike other social media platforms. Users may effortlessly add their favorite music and unique effects to their recorded films, thanks to the application's simple user interface.

11.2.3 MEDIA AND TECHNOLOGY NEED

Collaboration through cocreation in educational settings has been shown to be extremely beneficial for collaborative learning in these settings as well.

Students' learning and social media have been the subject of numerous studies. Web 2.0 tools allow students to collaborate on projects and information, according to Crook,[13] for example. Peer-to-peer assessment of assignments such as group projects can be improved by using social media, blogs, wikis, and online forums.

As an alternative to using Rice's[54] open-access, self-contained wiki to organize student research findings, students could use an open-access, self-contained wiki to gather data without restricting their research area. RSS documents could then be created to monitor changes in internet forums and podcasting sites regarding the subject matter. They can focus on researching information trends and building wiki connections and browser feeds to frequently discussed sites with this new, nontraditional data. By allowing students to publish their work and leave comments on how to improve it on a Wiki, these tools can reduce the amount of time spent delegating tasks and discussing writing strategies. Microsoft Teams, a cloud-based app that combines messaging, file sharing, and cocreation into a single app, has recently been developed to facilitate collaboration.[38]

Increased material, reduced coordination time, and the ability to work in groups are just a few advantages that social media can provide for educational collaborative production. However, social media platforms have been found to be useful in facilitating knowledge production and capture, as well as in sharing that knowledge.[13] A sense of community is also fostered among the students who work with Bharucha.[7] In addition, it was found that the use of Web 2.0 technology had a positive impact on student achievement in general.[24] Teachers are encouraged to incorporate social media tools into the classroom in order to keep up with the latest trend and improve learning outcomes for millennial students, whose lifestyles have become increasingly efficient and personalized, thanks to the use of social media tools.[39]

There isn't enough research on the effects of social media technologies on teaching and learning,[39] even though practitioners are experimenting with them.[24,39] According to Seaman and Tinti-Kane,[59] around 40% of teachers and teaching staff in the United States use social media tools in their lessons.

"Introductory" social media platforms like Facebook and Twitter have thrived because they are easier to use than blogs and wikis, according to Garca-Martn and Garca-Sánchez[19] (p 105). In addition, Table 11.1 summarizes the sharing capabilities of a variety of social media applications. This list is by no means complete; we list it based on the famous social media sites that have been accessed by people around the world.

TABLE 11.1 Social Media and Technology Tools Involved.[10]

	Major practices/activities involved	Tools
Users create content or create content together.	(1) Create content from various sources that can be accessed easily and is general in nature for public (2) Uploading content or sharing other content or can also be accessed by accessing content belonging to other people that can be adopted (3). Take part in creating content while using it (modifying) (4) Working in small groups to create works and collaborate on their production.	SNS: 1. Facebook, Twitter, LinkedIn, and Instagram YouTube Website/Wiki Blogspot Microsoft, Google Classroom, and LMS (Moodles) Forum: Zoom, Google Meet, WhatsApp, and Messenger.
Sharing and tagging	(1) Uploading and publishing, sharing content and editing the content. (2) Like and comment, blogging, and message (3) Tagging and sharing into the private profile and social network tagging (4) Screenshot or downloading	1. Social sharing: YouTube, Reels, Instagram 2. File sharing: WhatsApp, Web blog, and Padlet 3. Facebook, Twitter, and Instagram
Social community/ Group community	Socializing Discussing E-mobilizing E-message collaborating	Facebook WhatsApp Messenger Weblog Google sites

According to Petersen,[52] firms that operate social media tools reap all of the financial rewards while users—the losers—receive nothing. UGC has been renamed "loser-generated material" because of this belief. Under the terms of service, social media sites grant the firm a license to distribute, use, modify, duplicate, and display UGC, even though individuals retain ownership of the content (Facebook, YouTube, etc.). As the popularity and revenue generated by the site's users increased, so did the site's market value, which ultimately led to high-priced corporate acquisitions.

When it comes to internships, Deng and Tavares[14] claim that students' sense of community and social presence on Facebook led to high levels of engagement, ranging from casual conversation to intellectual discussions and problem-solving. With LinkedIn, users can establish professional career networks, join relevant professional organizations, and even follow "influencers" who are well-known business leaders.[11]

As a result, this chapter investigates the use of SOCMED and technology instructional design in EFL classes and finds out if SOCMED and technology instructional design are used for EFL learning and teaching in higher education.

11.3 METHODOLOGY

The study examined the teaching and learning experiences of students in higher education learning through ICT WhaYoZo (WhatsApp, YouTube, and Zoom) during the COVID-19 epidemic. It adopted a qualitative study approach because it enables the exploration of contextual circumstances and the realization of phenomena in a larger sense, hence facilitating comprehension and critical interpretation.[40] Additionally, the researchers used a phenomenological technique to help them explain and comprehend the individuals' experiences from their perspectives.[32]

The goal of the survey was to elicit instances of how students and instructors utilize e-learning tools and ICT to accomplish certain tasks during the educational process. These findings and participant input are included in the development of a blended learning model and the testing of the format of novel interaction patterns, as well as the development of creative teaching tactics for online learning at universities.

Additionally, the study collected 43 students' perceptions of those who taught during the COVID-19 epidemic. We sent a questionnaire using Google Forms after the learning and teaching process using ICT. Purposive sampling was used to pick the sample students. Purposive sampling is a nonprobability selection strategy that focuses on specific features of a population of interest in order to answer research questions.[61,64] It is determined by the study's aims and the researchers' knowledge of the population being studied.[70] Purposive sampling was used to locate and choose possible participants for the study who represented a larger spectrum of cases.[12] Interviews were conducted with participants. Additionally, as shown in Table 11.2, the participants in this study came from a variety of disciplines, and had varying levels of learning experience over 3 years with ICT, and had learned during the COVID-19 epidemic.

TABLE 11.2 Participants' Background.

Participants	Discipline	Sex	
		F	M
2	Mathematics department	1	1
2	Indonesian language department	2	0
2	Farm department	4	0
2	Agrotechnology department	2	1
7	Agribusiness department	4	3
4	Technical information department	2	3
5	Information systems department	2	3
2	Communication studies department	1	1
11	Public administration department	2	5
2	Public health department	2	1
3	Islamic economics department	2	1
∑ 43		24	19

The WhaYoZo (WhatsApp, YouTube, and Zoom) learning circle in this research is described in Picture 11.1 below:

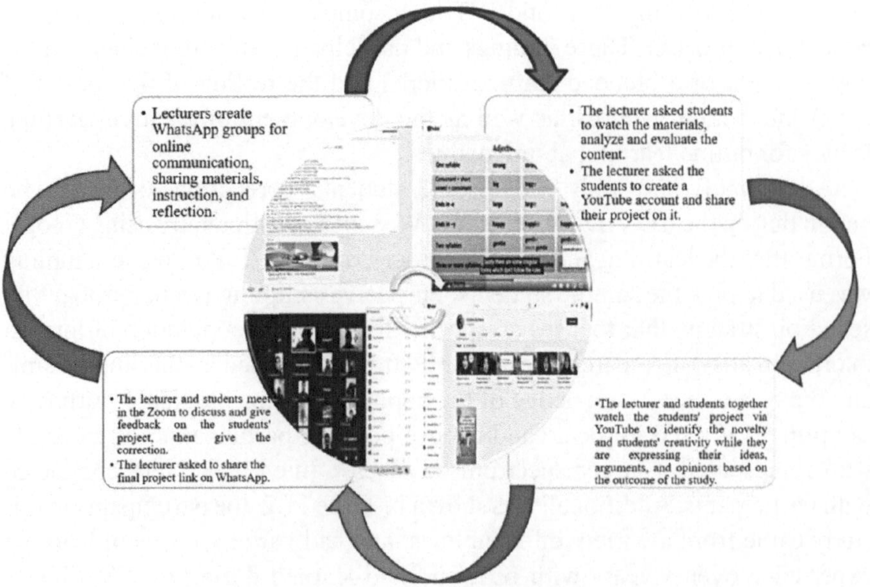

PICTURE 11.1 WhaYoZo (WhatsApp, YouTube, and Zoom) learning circle.

11.4 FINDINGS AND DISCUSSION

Transformational (educational) expectations are increasingly being linked to how digital technologies can be used to facilitate learning. Pedagogical change and innovation may be enabled through innovative uses of technology that promote student-controlled online learning activities, engage students, and provide a challenge for their own study. It's possible that we can assist kids in developing their own abilities to plan out and organize their own learning through tailored online activities with a fun and enjoyable learning approach. There is a growing trend among Indonesian universities to use student-owned technology to supplement classroom instruction in both formal and informal settings, with the goal of improving their students' digital literacy and job prospects. Integrating SOCMED is a recent example of the transformational change that the implementation plan for embedding mobile technologies in curriculum design represents in this respect. It empowers students in informal learning contexts like lab work to use their own smartphone devices in a range of practical e-learning settings that will be described in this part.

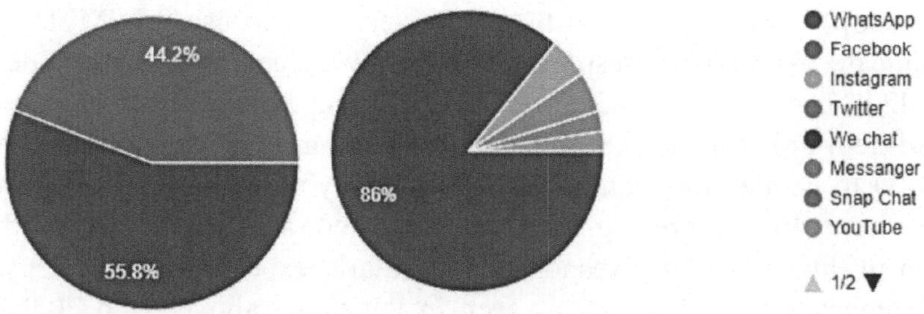

FIGURE 11.1 Student demographics.

In Figure 11.1, 19 were men and 24 were women among the survey's 43 participants. The participants in this study are students of English, ranging in age from 20 to 25 years old. None of the participants in the study were able to solve technological challenges because they lacked technical literacy, and they used SOCMED to interact mostly by using WhatsApp. According to 86% of respondents, 4.7% of respondents interact via Instagram, 2.3% via Twitter, and 2.3% via YouTube.

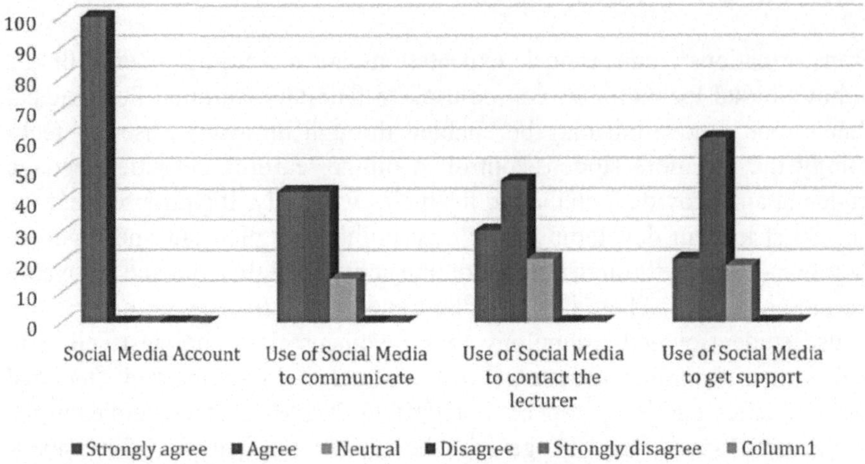

FIGURE 11.2 The students' use of social media in learning.

The students who used SOCMED in e-learning in Figure 11.2 described positive feedback on the statement about social media accounts (100% respondents have social media accounts, and the most frequently used is WhatsApp). Also, their claim that they were able to use SOCMED to communicate (42.9%), 43 respondents strongly agreed with these statements, 42.9% agreed, and 14.3% were in a neutral position. With regard to the third item in Figure 11.2, respondents stated that they were able to use SOCMED to contact the lecturer, with 30.2% saying strongly agree with this statement, 46.5% agreeing, and 20.9% in a neutral position. The fourth question in this poll asks if respondents regularly experience technology. The frequency rate is medium, as seen in the graph above, with 20.9% strongly agreeing. Despite this, 18.6% say they are "neutral," meaning they use technology occasionally.

During this section of the study, participants were quizzed on their understanding of the perceived usefulness of social media. These four questions include the usefulness of SOCMED in their learning, not only for interaction. To ensure that this is accurate, it should include a link to the source of the information. Figure 11.3 depicts the situation.

According to the frequency displayed in Figure 11.3, the study respondents agreed with the assertion that SOCMED can be used to get support for homework or assignments, 33.3% strongly agree with the statement, and 50% say they agree with the statement. There are, however, 16.7% of respondents who are still unsure about how SOCMED can be managed

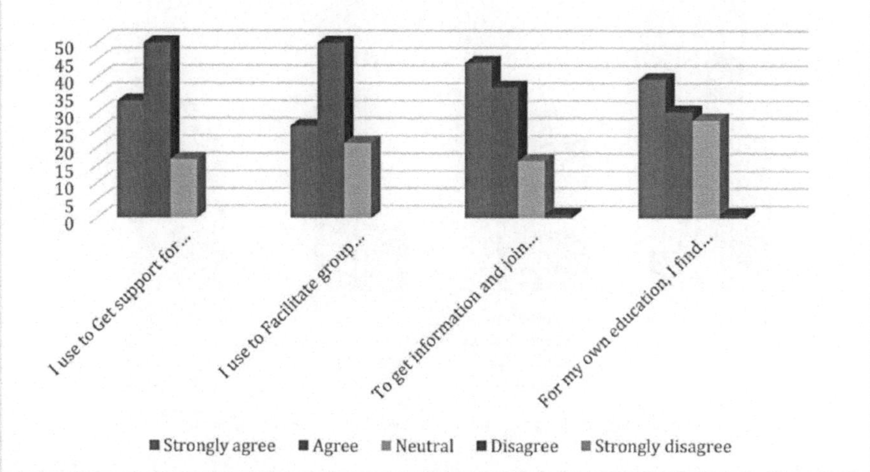

FIGURE 11.3 Perceived usefulness of the SOCMED.

as a learning tool. On the other hand, there is no respondent appraising, disagreeing, and strongly disagreeing. In the next statement, they were supported to facilitate group tasks, projects, and share projects. A total of 26.2% of the study participants feel they are able to use SOCMED to support their task or project, as stated in statement two. There are 50% agreeing and 21.4% of respondents in the "neutral" viewpoint. Fortunately, all responses to this question have the option of tailoring their learning to the students' varying levels of comprehension. When asked if they can use SOCMED to get information and join beyond-university/college projects/activities (Q3), 44.2% of respondents said they can do it very well, while 37.2% said they can do so fairly well. Only 16.3% of those polled said they could or couldn't adjust their SOCMED to get information. In contrast, 1% of those who took part in the survey can be said to be incapable of adapting SOCMED as a media and technology instructional tool for learning. The usefulness of SOCMED for students' own education and whether students find social media to be a useful tool is the following poll question (Q4). According to the graph, 39.5% of respondents strongly agreed with the statement, 30.2% agreed, while just 7% (1%) were unsure.

The primary focus of this session is on whether or not the participants in the study have an understanding of technology, the usefulness of SOCMED, and its content. Figure 11.4 depicts the four components of knowledge. When asked about their familiarity with students' common understandings of the power of SOCMED (Q1), 32.2 and 46.9% of respondents agreed (strongly

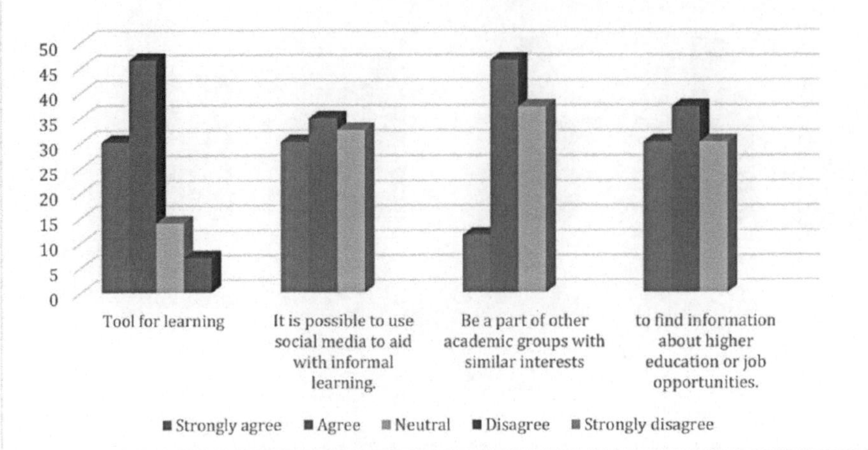

FIGURE 11.4 SOCMED can be a powerful tool.

agree and agree with positions) with the assertion of SOCMED as a powerful tool for better learning. Only 14% of them are still undecided. It's interesting to note that none of the respondents were unfamiliar with SOCMED, which indicates that SOCMED is well-versed with the students' learning needs.

Knowledge of it makes it possible to use social media to aid with informal learning statements on the survey. According to these statistics, 30.2% of respondents strongly agree and 34.9% agree with the statement. A total of 32.6% of respondents chose the neutral option, and no one said they disagreed with the premise. This study has found that all of the participants are able to manage their SOCMED in a positive manner. Research shows that all participants can learn`English using SOCMED, technology, and teaching techniques. A total of 11.6% of respondents strongly agree, and 46.5% of respondents agree that they can be a part of other academic groups with similar interests through SOCMED.

Research subjects believe they can learn better when lecturers mix their English courses with technology instruction in a way that works. The results showed that all of them are confident in their ability to interact with SOCMED in English courses as instructional design. When it comes to the capacity to pick the technology that enhances students' English skills, confidence in speaking, reducing anxiety, self-directed learning, autonomy, and what students learn in a classroom, the findings in this study bear this out. Of those polled, they also found that SOCMED is a powerful tool to find information about higher education or job opportunities for their next career. On this statement, 30.2% strongly agreed, 37.2 agreed, 0% disagreed, and 30.27% were neutral.

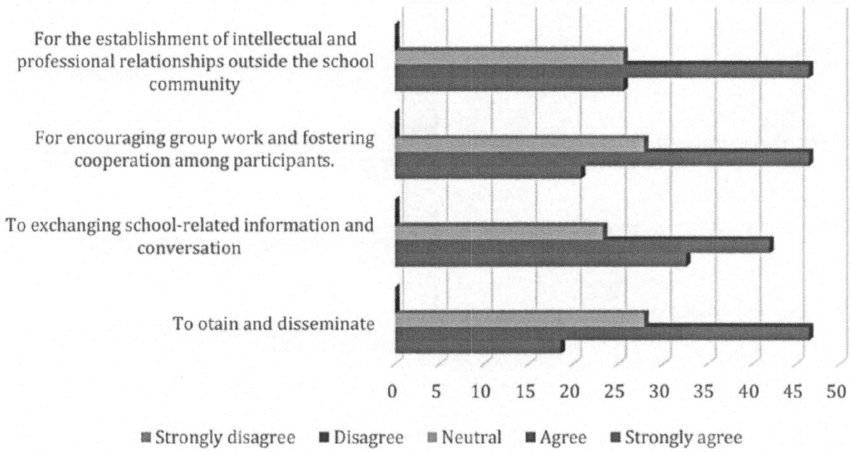

FIGURE 11.5 SOCMED as educational tool and material.

Using the following questions in Figure 11.5, the percentage of perceived SOCMED as an educational tool and material for learning English is calculated as follows: When it comes to my English classes, employing SOCMED-WhaYoZo-based platforms (WhatsApp, YouTube, and Zoom) can help me learn more quickly and also facilitate the establishment of intellectual and professional relationships outside the school community. According to the findings, 25.6% of respondents strongly agreed, 46.5% agreed, and 25.6% were neutral or disagreed. They believe SOCMED's encouraging group work and fostering cooperation among participants may help me enhance my English course performance. It may also help students increase their English course efficiency, according to the second question. According to the results, 20.9% strongly agree, 45.5% agree, and 20.5% were neutral. The third question, SOCMED as an educational tool to exchange school-related information and conversation, had 32% of respondents from 43 choosing to strongly agree, 41.9% agree, and 23.3% were neutral on this statement. The final question is whether or not SOCMED can boost efforts to obtain and disseminate the materials. According to the findings (0%), 18.6% of respondents indicated strong agreement, 46.5% expressed agreement, 27.9% expressed neutrality, and 0% expressed disagreement. Using SOCMED as media and technology instruction or e-learning will make the educational process and English courses simpler.

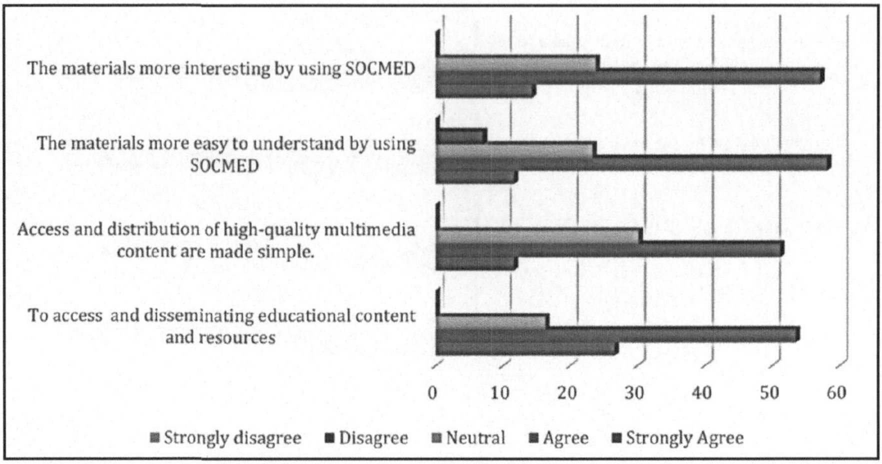

FIGURE 11.6 The use of SOCMED-based content materials.

In Figure 11.6 for this survey, we need to answer four questions: the use of SOCMED as teaching materials is a great way to incorporate technology into English classes and also to access and disseminate educational content and resources. According to the findings, 26.6% of respondents strongly agreed, 53.5% agreed, and 16.3% were neutral. In regard to the second topic, the access and distribution of high-quality multimedia content are made simple. There were 11.6% of respondents who said they strongly agreed, 51.2% who said they agreed, and 30.2% who said they were neutral.

The third statement is: Is it a good idea to incorporate materials that are easier to understand by using SOCMED into English classes? The study found that 11.6% of respondents strongly agreed with the category, 25 people agreed, 58.1% agreed with the category, 30.1% were neutral, and 7% disagreed. In the last question asking about the materials being more interesting by using SOCMED, 14.3% of respondents strongly believed, 57.1% agreed, and 23% were in a neutral position; none of them disagreed or strongly disagreed.

It was found that study subjects were in agreement with the statement, based on the frequency given in Figure 11.7, in question one regarding how the integration of social media and technology into learning is helping students learn better. A total of 32.6% strongly agree with the statement, and 53.5% say they agree with the statement. There are, however, 14% of respondents who are still undecided. The second statement, asking about the students' attitudes and behaviors toward the integration of social media and technology into learning, helps them understand the materials. According

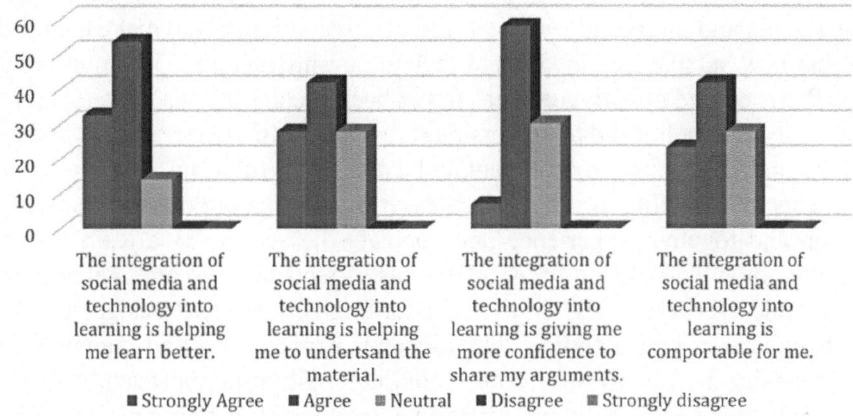

FIGURE 11.7 The students' perception on the integration of SOCMED and technology in attitude and behavior.

to the graph, 27% of respondents strongly agree with the statement, while 41.9% agree and 27% are neutral. None of them chose to disagree or strongly disagree. Almost on the same line, the incorporation of social media and technology into learning is giving students more confidence to share their points of view. Of those, 7% strongly agree, 41.9% agree, and 30.2% are neutral. The integration of social media and technology as teaching materials into learning is manageable for students. A total of 23.3% of respondents strongly agree with this position, 41.9% agree, and 27.9% take a neutral stance.

In Asia, the majority of higher education institutions used a mix of traditional and blended learning approaches prior to the arrival of COVID-19. Nevertheless, the proliferation of COVID-19 necessitated that many educational institutions either cease their academic activity or adapt to online instruction. Indonesian higher education institutions resumed their academic activity online after a brief halt period. Similar to that of the instance university, most of the faculty were unfamiliar with teaching online when the institution shifted to online learning. Students and professors alike had to learn and develop new sets of abilities, capabilities, and venues. The moment has come to review the efficiency and success of these efforts after numerous months (totaling over 2 years currently). Questions about the quality of online teaching and student happiness are critical because any unsolved concerns might have long-term ramifications for the structure of the higher education industry and the future shape of the programs produced and delivered in the sector.

This chapter examines how students view SOCMED and technology instruction and offers suggestions for how universities should make the most of the new standard of media and technology instruction. The importance of efficiency and effectiveness has never been greater because this is now a necessity. Our belief is that universities are well-positioned to take advantage of the online learning environment and extend their offerings. Alternatively, students may be able to join the professor in a classroom or via WhatsApp group and together experience and cocreate the session as a team group. Greater knowledge of the "new" client base—students who are interested in online and hybrid learning—is essential for English courses to design, launch, and implement real activities that suit their expectations and demands. If a university decides to offer both an online and hybrid approach, it might have an impact on both their current outcomes and their future growth. This chapter expands the literature on SOCMED and technology instruction by incorporating the perspectives of university students from different study backgrounds.

It was discovered in a study conducted by Zgheib and Dabbagh (2013) that lecturers and students alike saw social media in the classroom as an educational tool that could introduce students to new technologies while also improving their academic performance. A study by Haidari et al. (2020) found that wikis and social media were useful in helping students improve their writing and general skills. Additionally, the study found that students' abilities to communicate, work in groups, collaborate, and think critically all increased as a result of their exposure to social media. It means that the students utilize social media for more than just communication. They use it for educational purposes as well. A study by Rodriguez[55] found that lecturers incorporated social media into their instruction in order to increase student engagement and motivation in the classroom. Students, according to Li,[34] utilize social media to spread knowledge and get educational resources. In order to keep up with their class assignments and projects, the respondents turned to social media for information. According to Esteve et al.,[17] social media is widely used in teaching and helps instructors stay up-to-date and learn more about their students' progress.

The effectiveness of SOCMED and technology instruction on education versus face-to-face learning has been tested and evaluated in a variety of ways in different researches. Student results, such as test scores and grades; student attitudes about learning; and general student satisfaction with online learning, all fall into three main categories and are also identified by Whitaker et al.;[69] however, they use somewhat different language. First, they look at how technology can be used to improve student learning outcomes, followed by comparisons of in-class, online, and blended course outcomes. Finally, they look at whether online education is suitable for various student types.

Besides the conditions in which online courses are taught and the way programs are specifically customized for the online environment, we also looked at the characteristics of students. A student's personal characteristics and preferences are also crucial in choosing their preferred method, media, materials, activities, and technology instructions. A variety of demographic and psychological characteristics must be taken into account when constructing a profile of the typical student in an online education program.

As lecturers, we can choose media, tools, or technology in designing the materials and instruction. In this study, through SOCMED, we see that students often explain their misunderstanding of the materials because of a lack of sources. Our research shows that students who were less involved in online or hybrid learning environments before, while providing teaching materials through SOCMED in the pandemic when learning remotely, started to increase their activity in learning. Lecturers are faced with the challenge of verifying to what extent the SOCMED engages students to reduce passive activities.

This chapter looked at how social media may be used to enhance EFL technology instruction at the university level in Indonesia. It also looked at how students and teachers of English as a foreign language like to use social media as learning and teaching material for English courses and their experience in learning what it's like to study a native language. As a consequence of the study's findings, Facebook, WhatsApp, and YouTube emerged as three of the most popular social media platforms for EFL students and teachers. In this study, the students learned more from YouTube learning than from Facebook, and for meeting online, they preferred Zoom over the WhatsApp call group. As a counterargument, this study finds that YouTube was one of the least frequently utilized social networking sites in EFL learning and teaching, according to Tantarangsee et al.[63] As a result of this research, it was discovered that Facebook, Twitter, Messenger, and Instagram were not used in Indonesian EFL learning and instruction.

Learning and education have been transformed by social media, which has been widely used for a variety of objectives. The study found that teachers of English as a second language (ESL) rely heavily on social media. The fact that students and professors communicate via social media was also revealed, as were the professors' talents in utilizing social media in their teaching. Social media was found to be a time-saver by the majority of respondents, and they credited it with improving their university teaching. Social media has been found to be an excellent tool for teaching and studying ESL. Around 75% of participants, on the other hand, stated that both instructors and students were suffering from eye difficulties as a result of their excessive use of social media. Researchers Haand and Shuwang (2020) observed that both instructors and students had health difficulties as a result of their excessive usage of social media.

According to the findings, students utilize social media to supplement their classroom instruction. Most students reported that social media helped them stay up to speed on course information and access materials and academic resources more quickly. The students' responses, writing, reading, speaking, listening, and learning habits have been enhanced by the use of SOCMED. It was found that students were able to learn about class schedule changes, share materials, create YouTube accounts, and share their tasks, answers, conversations, opinions and arguments, and other details via social media, making it easier for them to study and understand the courses at their own pace. A recent study conducted by Muftah[43] and Li[34] shows student comfort with online class design, organization, the interaction between students and professors, quality material, as well as overall online class experience. Furthermore, Haque and Al-Salem (2019) have also demonstrated that students can benefit from using social media to learn foreign languages, acquire resources, and enhance their abilities. Al-Taweel et al. (2010) indicated that students made extensive use of social media to disseminate knowledge and access educational resources. Social media was also a source of program and course updates for the study's participants.

The majority of students in this study believed that social media was a familiar and useful tool for enhancing their academic performance, such as confidence in expressing their ideas, helping them learn better ways to express their English in writing, speaking, reading, listening, and arguing, and also for getting support and self-reflection during online learning. The university should take full advantage of this finding and its impact to create a learning environment that is more fun and comfortable and reduces anxiety in learning English. It also develops English courses for native speakers and helps regulate students' use of social media in the classroom. For universities in Indonesia developing SOCMED and technology instruction, it appears that students and lecturers must work together to determine the best methods to use technology, such as social media, in education in unusual circumstances. SOCMED and technology instructions in this study were described by giving the training to the students and also asking them to create their technology accounts for their English courses, and they were asked to upload their English projects after learning activities.

11.5 CONCLUSION

Social media has been widely utilized by Indonesian university lecturers and students, but no research has been done to examine how students and teachers of English as a foreign language use SOCMED as courses and technology instruction in teaching. This study fills a void by analyzing the use of media in EFL teaching and learning, as well as the frequency and preference of the use of social media in EFL learning and teaching, thereby completing the research. Social media was found to be widely used in EFL learning and teaching. According to the findings of the study, WhaYoZo (WhatsApp, YouTube, etc.) was the most popular social networking app utilized in Indonesian higher education for teaching and studying EFL. According to this study, most students and lecturers at universities use various types of social media in their teaching and learning activities, including social media for communication, reading, sharing tasks, sharing video projects, sharing link content, expressing ideas, arguments, creativity, materials design, course information, sharing projects, online meetings, presentation tools, and self-reflection.

The findings of this investigation have consequences for both the real world and the theoretical realm. Indonesian higher education officials should develop a freedom to learn strategy to increase university facilities and encourage people to learn and teach via social media besides WhaYoZo (WhatsApp, YouTube, and Zoom) because resources and technical capabilities are limited. Researchers in the future may use a larger sample and mixed methods studies to target university students and instructors from both public and private institutions. The use of social media and technological instruction design in teaching for the next scenario in EFL education may be the subject of future research.

Educators must be prepared for a paradigm shift from traditional education to pedagogy boosted by modern technologies in order to incorporate the good findings of the research. In order to better understand how digital media usage affects student attitude, reduce anxiety in retelling the materials, have confidence in expressing their ideas, feelings, problem-solving, collaboration, and critical thinking that come with it, and other detrimental effects on self-directed learning, self-determined learning, and cognitive behavior, more study is needed. When it comes to using social media, this study serves as a starting point and should serve as a catalyst for further research.

KEYWORDS

- **social media**
- **online**
- **technology**
- **instructional design**
- **education**

REFERENCES

1. Affouneh, S.; Salha, S.; Khlaif, Z. N. Designing Quality e-Learning Environments for Emergency Remote Teaching in Coronavirus Crisis. *Interdiscip. J. Virtual Learn. Med. Sci.* **2020,** *11* (2), 135–137.
2. Akcil, U.; Uzunboylu, H.; Kinik, E. Integration of Technology to Learning-Teaching Processes and Google Workspace Tools: A Literature Review. *Sustainability* **2021,** *13* (9), 5018.
3. Akhmedov, B. A. Innovative Cluster Model for Improving the Quality of Education. *Acad. Res. Educ. Sci.* **2021,** *2* (3).
4. Al-Rahmi, A. M.; Al-Rahmi, W. M.; Alturki, U.; Aldraiweesh, A.; Almutairy, S.; Al-Adwan, A. S. Exploring the Factors Affecting Mobile Learning for Sustainability in Higher Education. *Sustainability* **2021,** *13* (14), 7893.
5. Arkorful, V.; Barfi, K. A.; Aboagye, I. K. Integration of Information and Communication Technology in Teaching: Initial Perspectives of Senior High School Teachers in Ghana. *Educ. Inform. Technol.* **2021,** *26* (4), 3771–3787.
6. Bailey, D. R.; Lee, A. R. Learning from Experience in the Midst of Covid-19: Benefits, Challenges, and Strategies in Online Teaching. *Comput. Assist. Lang. Learn. Electron. J.* **2020,** *21* (2), 178–198.
7. Bharucha, J. Exploring Education-Related Use of Social Media: Business Students Perspectives in a Changing India. *Educ. Train.* **2018,** *60* (2), 198–212.
8. Brown, M.; Hughes, H.; Keppell, M.; Hard, N.; Smith, L. Stories from Students in Their First Semester of Distance Learning. *Int. Rev. Res. Open Distrib. Learn.* **2015,** *16* (4), 1–17.
9. Cabero Almenara, J.; Morales Lozano, J. A.; Barroso Osuna, J. M.; Fernández Batanero, J. M.; Romero Tena, R.; Román Graván, P.; … Ballesteros Regaña, C. Análisis de Centros de Recursos de Producción de las TIC de las Universidades Españolas. *Rev. Educ.* **2010,** *351,* 237–257.

10. Chu, S. K. W. *Social Media Tools in Experiential Internship Learning*; Springer: Berlin, Germany, 2020.
11. Cooper, B.; Naatus, M. K. LinkedIn as a Learning Tool in Business Education. *Am. J. Bus. Educ.* **2014,** *7* (4), 299–306.
12. Creswell, J. W.; Poth, C. N. *Qualitative Inquiry and Research Design: Choosing Among Five Approaches: Sage Publications*; Thousand Oaks, CA, 2007.
13. Crook, C.; Fisher, T.; Graber, R.; Harrison, C.; Lewin, C.; Cummings, J.; ... Sharples, M. Implementing Web 2.0 in Secondary Schools: Impacts, Barriers and Issues, 2008.
14. Deng, L.; Tavares, N. J. From Moodle to Facebook: Exploring students' Motivation and Experiences in Online Communities. *Comput. Educ.* **2013,** *68,* 167–176.
15. Dziuban, C.; Graham, C. R.; Moskal, P. D.; Norberg, A.; Sicilia, N. Blended Learning: The New Normal and Emerging Technologies. *Int. J. Educ. Technol. Higher Educ.* **2018,** *15* (1), 1–16.
16. Echeverría Sáenz, A. C. Use of TIC (Communication and Information Technologies) in University Teaching: Opinion of Special Education Professors. *Actual. Investig. Educ.* **2014,** *14* (3), 272–297.
17. Esteve Del Valle, M.; Gruzd, A.; Haythornthwaite, C.; Paulin, D.; Gilbert, S. In *Social Media in Educational Practice: Faculty Present and Future Use of Social Media in Teaching,* Proceedings of the 50th Hawaii International Conference on System Sciences, 2017.
18. Ferro Soto, C.; Martínez Senra, A. I.; Otero Neira, M. C. Ventajas del uso de Las TICs en el Proceso de Enseñanza-Aprendizaje Desde la Óptica de Los Docentes Universitarios Españoles. *Edutec: Revista Electrónica de Tecnología Educativa*, 2009.
19. García-Martín, J.; García-Sánchez, J. N. Patterns of Web 2.0 Tool Use Among Young Spanish People. *Comput. Educ.* **2013,** *67,* 105–120.
20. Ghavifekr, S.; Wong, S. Y. Technology Leadership in Malaysian Schools: The Way Forward to Education 4.0–ICT Utilization and Digital Transformation. *Int. J. Asian Bus. Inform. Manag.* **2022,** *13* (2), 1–18.
21. Ghobrini, R. E. A.; Benzert, F. Z.; Balas, M. Educationalizing Instagram for Virtual Instruction in COVID-19: A Pragmatic Framework. *Int. J. Web Based Learn. Teach. Technol.* **2022,** *17* (6), 1–16.
22. Hafeez, M.; Kazmi, Q. A.; Tahira, F.; Hussain, M. Z.; Ahmad, S.; Yasmeen, A.; ... Saqi, M. I. Impact of School Enrolment Size on Student's Achievements. *Indones. J. Basic Educ.* **2020,** *3* (1), 17–21.
23. Hermans, R.; Tondeur, J.; Van Braak, J.; Valcke, M. The Impact of Primary School Teachers' Educational Beliefs on the Classroom Use of Computers. *Comput.Educ. 51* (4), 1499–1509.
24. Hew, K. F.; Cheung, W. S. Use of Web 2.0 Technologies in K-12 and Higher Education: The Search for Evidence-Based Practice. *Educ. Res. Rev.* **2013,** *9,* 47–64.
25. Hrastinski, S. What do We Mean by Blended Learning? *TechTrends* **2019,** *63* (5), 564–569.
26. Huszár, F.; Ktena, S. I.; O'Brien, C.; Belli, L.; Schlaikjer, A.; Hardt, M. Algorithmic Amplification of Politics on Twitter. *Proc. Natl Acad. Sci.* **2022,** *119* (1).
27. Karimi, H.; Torphy, K. T.; Derr, T.; Frank, K. A.; Tang, J. In *Characterizing Teacher Connections in Online Social Media: A Case Study on Pinterest*, Proceedings of the Seventh ACM Conference on Learning@ Scale, 2020, August; pp 249–252.

28. Kebritchi, M.; Lipschuetz, A.; Santiague, L. Issues and Challenges for Teaching Successful Online Courses in Higher Education: A Literature Review. *J. Educ. Technol. Syst.* 2017, *46* (1), 4–29.
29. Khan, T. M. Use of Social Media and WhatsApp to Conduct Teaching Activities During the COVID-19 Lockdown in Pakistan.*Int. J. Pharm. Pract.* **2021,** *29* (1), 90–90.
30. Kramarenko, T.; Bondar, K.; Shestopalova, O. The ICT Usage in Teaching Mathematics to Students with Special Educational Needs. *J. Phys. Conf. Ser.* **2021,** *1840* (1), 012009.
31. Kumbar, S. S.; Bidnurkar, A.; Tamhane, K. N.; Kumbar, R. B. Case Study on Need of ICT in Teaching Learning Process. *J. Eng. .ation Transform.* **2018,** *32* (1), 7–11.
32. Lambert, M. *Practical Research Methods in Education*; Routledge: New York, 2019.
33. Lareki, A.; de Morentin, J. I. M.; Amenabar, N. Towards an Efficient Training of University Faculty on ICTs. *Comput. Educ.* **2010,** *54* (2), 491–497.
34. Li, V. Social Media in English Language Teaching and Learning. *Int. J. Learn. Teach.* **2017,** *3* (2), 148–153.
35. López-Carril, S.; Anagnostopoulos, C.; Parganas, P. Social Media in Sport Management Education: Introducing LinkedIn. *J. Hosp. Leis. Sports Tour. Educ.* **2020,** *27,* 100262.
36. Lyken-Segosebe, D.; Gamariel, G.; Bagai, K. Boundary Crossing for Education Continuity: Exploring WhatsApp's Potential in Botswana During and Beyond the Pandemic. *iJIM* **2022,** *16* (02), 59.
37. Manca, S.; Ranieri, M. Facebook and the Others. Potentials and Obstacles of Social Media for Teaching in Higher Education. *Comput. Educ.* **2016,** *95,* 216–230.
38. Martin, L.; Tapp, D. Teaching with Teams: An Introduction to Teaching An Undergraduate Law Module Using Microsoft Teams. *Innov. Pract. Higher Educ.* **2019,** *3* (3).
39. McHaney, R. *The New Digital Shoreline: How Web 2.0 and Millennials are Revolutionizing Higher Education*; Stylus Publishing, LLC, 2012.
40. McNabb, D. E. *Research Methods in Public Administration and Nonprofit Management*; Routledge, 2017.
41. Mhlana, S.; Chipangura, B.; Twinomurinzi, H. In *Conceptualizing Factors that Influence Learners' Intention to Adopt ICT for Learning in Rural Schools in Developing Countries*, Proceedings of Sixth International Congress on Information and Communication Technology; Springer: Singapore, 2022, pp 391–401.
42. Milićević, V.; Denić, N.; Milićević, Z.; Arsić, L.; Spasić-Stojković, M.; Petković, D.; ... Jovanović, A. E-learning Perspectives in Higher Education Institutions. *Technol. Forecast. Soc. Change* **2021,** *166,* 120618.
43. Muftah, M. Impact of Social Media on Learning English Language during the COVID-19 Pandemic. *PSU Res. Rev.* 2022.
44. Muñoz-Miralles, R.; Ortega-González, R.; López-Morón, M. R.; Batalla-Martínez, C.; Manresa, J. M.; Montellà-Jordana, N.; … Torán-Monserrat, P. The Problematic Use of Information and Communication Technologies (ICT) in Adolescents by the Cross Sectional JOITIC Study. *BMC Pediatr.* **2016,** *16* (1), 1–11.
45. Muthmainnah, N. A.; Galal, M.; Varghese, K. J.; Del Castillo, F.; Ghofur, A. The Students' Needs in Developing EFL Materials ICT Based. *OKARA* **2021,** *15* (2), 235.
46. Nikolić, V.; Petković, D.; Denić, N.; Milovančević, M.; Gavrilović, S. Appraisal and Review of e-Learning and ICT Systems in Teaching Process. *Phys. A: Stat. Mech. Appl.* **2019,** *513,* 456–464.
47. Nikonov, E.; Shamis, E. *Generations Theory Basics*; Synergy: Moscow, 2016.

48. Noori, A. Q.; Orfan, S. N.; Akramy, S. A.; Hashemi, A. The Use of Social Media in EFL Learning and Teaching in Higher Education of Afghanistan. *Cogent. Soc. Sci.* **2022,** *8* (1), 2027613.
49. Núñez-Canal, M.; de Obesso, M. D. L. M.; Pérez-Rivero, C. A. New Challenges in Higher Education: A Study of the Digital Competence of Educators in Covid Times. *Technol. Forecast. Soc. Change* **2022,** *174,* 121270.
50. Orellana, N.; Bo, R.; Belloch, C.; Aliaga, F. Estilos de Aprendizaje y Utilización de las TIC en la Enseñanza Superior. 2010.
51. Patru, M.; Balaji, V. Making Sense of MOOCs: A Guide for Policy-Makers in Developing Countries, 2016.
52. Petersen, S. M. Loser Generated Content: From Participation to Exploitation. *First Monday*, 2008.
53. Picciano, A. G. Theories and Frameworks for Online Education: Seeking an Integrated Model. In *A Guide to Administering Distance Learning*; Brill, 2021; pp 79–103.
54. Rice, J. A. Devising Collective Knowledges for the Technical Writing Classroom: A Course-Based Approach to Using Web 2.0 Writing Technologies in Collaborative Work. *IEEE Trans. Prof. Commun.* **2009,** *52* (3), 303–315.
55. Rodriguez, J. E. Social Media Use in Higher Education: Key Areas to Consider for Educators, 2011.
56. Samoylenko, N.; Zharko, L.; Glotova, A. Designing Online Learning Environment: ICT Tools and Teaching Strategies. *Athens J. Educ.* **2022,** *9* (1), 49–62.
57. Sayaf, A. M.; Alamri, M. M.; Alqahtani, M. A.; Al-Rahmi, W. M. Information and Communications Technology Used in Higher Education: An Empirical Study on Digital Learning as Sustainability. *Sustainability* **2021,** *13* (13), 7074.
58. Sayaf, A. M.; Alamri, M. M.; Alqahtani, M. A.; Alrahmi, W. M. Factors Influencing University Students' Adoption of Digital Learning Technology in Teaching and Learning. *Sustainability* **2022,** *14* (1), 493.
59. Seaman, J.; Tinti-Kane, H. *Social Media for Teaching and Learning*; Pearson Learning Systems: London, 2013.
60. Sharifzadeh, M.; Agah, J.; Khosravi, A.; Samadi, S.; Sani, S. J. D. Collaborative Method Consisting Lecture, Problem-Based Learning and Weblog for Clinical Courses of Medical Students in Comparison with Lecture Method. *J. Educ. Health Promot.* **2021,** *10.*
61. Silverman, D. *Doing Qualitative Research*, 4th ed.; SAGE Publications, Inc.: Los Angeles, CA, USA, 2013.
62. Tahat, K. M.; Al-Sarayrah, W.; Salloum, S. A.; Habes, M.; Ali, S. The Influence of YouTube Videos on the Learning Experience of Disabled People During the COVID-19 Outbreak. In *Advances in Data Science and Intelligent Data Communication Technologies for COVID-19*; Springer: Cham, 2022; pp 239–252.
63. Tantarangsee, C.; Kosarassawadee, N.; Sukweses, A. The Use of Social Media in Teaching and Learning: A Case of SSRU's Faculty Members. *Int. J. Innovat. Manag. Technol.* **2017,** *8* (6), 471–476.
64. Teddlie, C.; Yu, F. Mixed Methods Sampling: A Typology with Examples. *J. Mixed Meth. Res.* **2007,** *1* (1), 77–100.
65. Todorovic, M.; Coyne, E.; Gopalan, V.; Oh, Y.; Landowski, L.; Barton, M. Twelve Tips for Using Facebook as a Learning Platform. *Med. Teach.* **2021,** *43* (11), 1261–1266.

66. Top, E.; Baser, D.; Akkus, R.; Akayoglu, S.; Gurer, M. D. Secondary School Teachers' Preferences in the Process of Individual Technology Mentoring. *Comput. Educ.* **2021,** *160,* 104030.
67. Tubin, D. Typology of ICT Implementation and Technology Applications. *Comput. Sch.* **2006,** *23* (1–2), 85–98.
68. Vogiatzis, D.; Charitonos, K.; Giaxoglou, K.; Lewis, T. Can WhatsApp Facilitate Interaction? A Case Study of Adult Language Learning. In *Open World Learning*; Routledge, 2022; pp 44–62.
69. Whitaker, J.; New, J. R.; Ireland, R. D. MOOCs and the Online Delivery of Business Education What's new? What's not? What now? *Acad. Manag. Learni. Educ.* **2016,** *15* (2), 345–365.
70. Wiersma, W.; Jurs, S. G. *Research Methods in Education: An Introduction*, 8th ed.; Pearson Education Asia Ltd.: Hong Kong; China Light Industry Press: Beijing, China, 2004.
71. Xodabande, I. The Effectiveness of Social Media Network Telegram in Teaching English Language Pronunciation to Iranian EFL Learners. *Cogent Educ.* **2017,** *4* (1), 1347081.
72. Yang, H. In *Secondary-School Students' Perspectives of Utilizing Tik Tok for English Learning in and Beyond the EFL Classroom*, 2020 3rd International Conference on Education Technology and Social Science (ETSS 2020), 2020; pp 163–183.
73. Yazon, J. M. O.; Mayer-Smith, J. A.; Redfield, R. J. Does the Medium Change the Message? The Impact of a Web-Based Genetics Course on University Students' Perspectives on Learning and Teaching. *Comput. Educ.* **2002,** *38* (1–3), 267–285.
74. Zhao, Y.; Tan, S. H.; Mishra, P. Teaching and Learning: Whose Computer is it? *J. Adolesc. Adult Literacy* **2000,** *44* (4), 348.
75. Zhu, C.; Xu, X.; Zhang, W.; Chen, J.; Evans, R. How Health Communication via Tik Tok Makes a Difference: A Content Analysis of Tik Tok Accounts Run by Chinese Provincial Health Committees. *Int. J. Environ. Res. Public Health* **2020,** *17* (1), 192.
76. Zourmpakis, A. I.; Papadakis, S.; Kalogiannakis, M. Education of Preschool and Elementary Teachers on the Use of Adaptive Gamification in Science Education. *Int. J. Technol. Enhanc. Learn.* **2022,** *14* (1), 1–16.

CHAPTER 12

Factors Affecting Placement Preparation in the COVID-19 Era: A Case Study

SHRIRAM RAGHUNATHAN[1], BOUDHAYAN BANDYOPADHYAY[2], and MANAS KUMAR MISHRA[1]

[1]*School of Computer Science and Engineering, VIT Bhopal University, Kothri Kalan, Madhya Pradesh, India*

[2]*Department of Biotechnology, School of Life Science and Biotechnology, Adamas University, Kolkata, India*

ABSTRACT

In 2020, all the sectors across the world faced huge turbulence due to COVID-19. Despite this disturbing situation, academic institutions have to continue training activities to ensure the students' future toward their dream jobs. The classroom activities were transformed into virtual ones with immediate effect. When a new system is forced to be introduced in place of the traditional system, it is envisaged that new challenges will come in different dimensions. Considering this scenario, we wanted to study the types of challenges the students may face from these newly introduced online classroom activities during their training for recruitment. Using several postonline classroom surveys related to web connectivity, this work focuses on the proper utilization of online course materials and gamified platforms and performance in online tests. The focus was on the correlation of the data with students' psychological determinants which can channel them to cope with this new situation. The data survey showed that the primary determinants are self-motivation which comes from their exam scores and teacher's effort. So far to our knowledge, there is no report related to placement training through online learning platforms. This study will help the academic institutions and

Global Digital Transformation and the Covid-19 Pandemic.
Rajdeep Chowdhury, K. Martin Sagayam, Neyara Radwan, Ahmad A. Elngar (Eds.)

vendors who are putting lots of effort into successful placement training and also the student communities who are looking for answers in this regard.

12.1 INTRODUCTION

Pandemic diseases have struck several times, causing the deaths of millions of people. Every time, humanity coped with the situation and we returned to daily equilibrium within short periods. COVID-19 has affected the balance and the daily pace of life. As the number of COVID-19-positive cases started increasing exponentially worldwide, the lockdown was implemented everywhere to reduce the curve. This situation affected every sector of society across the world, hampering the daily lives of the people and worsening the economic conditions of every nation worldwide. The students of the prefinal year in India prepare for their recruitment process.

Typically, B.Tech. students prepare for their recruitment intensively from January to July, and companies start visiting the campuses from July onward. In India, in 2020, the lockdown was declared in March. There was, however, no change in the recruitment dates. Thus, students' preparedness was severely hampered as they dealt with a lot of uncertainty throughout their placement process. Educators and placement officers tried their utmost to fill the void by adopting various online training strategies.

Online education before this was purely limited to webinars and never tried as a part of Engineering education in India. However, students were familiar with online courses and had learned to use them a lot. The difference between online and live online courses was the interactivity. Training for recruitment relies on having synchronous discussions, peer support, and a good set of interactions between the stakeholders. Moving online removes the personalization and peer support element and the challenge was daunting.

The training for recruitment is a result-driven analytical approach and is a subset of the broader training that happened online. The big difference is the focus on problem-solving skills, repeated evaluation, peer help in learning, and the impact of gamified approaches. The focus of training is not just understanding but the application of knowledge. At the same time, it is also a result-driven process wherein students learn by testing themselves over a period of time.

This work relies on a set of surveys conducted with students from April 2020 to the completion of their placement processes in May 2020 and shows a snapshot of students' issues and challenges. Participation in the surveys was purely voluntary.

This research highlights the challenges that learners face while attending online training from home. We believe that these challenges are not unique to any specific institution, but are a common problem faced by all institutions across the globe. The general finding of our report is that student motivation is the biggest factor in online learning. Motivation is directly correlated to the scores of the students and their participation in voluntary studies. There is a definitive impact of lean media and gamified approaches in learning interventions.

12.2 LITERATURE REVIEW

Agarwal and Kaushik focus on learning during COVID-19, concluding that motivation is an important factor in the pandemic era.[2] This is similar to the results published by Lall and Singh.[16] The study by Fansury et al.[6] focuses on the usage and influence of digital content, which can help in enhancing student learning. An interesting study was reported,[4] where graduate students in the doctoral program discussed the effect of pandemics on their academics.

In another study,[21] the perspective of teachers was surveyed and mixed responses were found. Some of the teachers accepted online teaching as a significant benefit, whereas others did not feel confident and comfortable about it. Izumi et al.[11] explain the importance of the stakeholders to adapt to emergency scenarios. From the survey across 65 universities, located in 29 countries, the major challenge found was the lack of preparedness for the pandemic and the practice related to it. Kaur et al.[13] showcased the challenges and opportunities that have come in this era.

Lean media can have a positive impact on knowledge-sharing when used for simple activities.[14] Lean media can also engage people in high-level group communication and better group outcomes.[15] There is evidence that being deliberate about media structures in technology-enhanced training is worthwhile.[5,17]

One of the biggest challenges during the initial days of COVID-19 was the uncertainty about the future in all aspects of educational learning. The worry about how learning will happen, how effective it will be, how it will be evaluated, and finally the impact of the process were the key factors.

Research work like Woywodt et al.[22] reflects this angst. While there has been various research works focused on all aspects of the teaching learning processes, the spotlight on graduate recruitment has been limited. In India, during March, when the lockdown was imposed students were preparing for their graduate recruitment processes that will traditionally begin in July in an offline mode. The sudden lockdowns and the subsequent migration to online learning meant that students had to calibrate their own studies and preparation in a new way.

Holt-White and Montacute[10] focus on the graduate recruitment aspect in July 2020 and focused on the impact on students. The primary concern was the loss of learning, lack of opportunities for young graduates, and reduced placement prospects especially in non-urban cities. This is an interesting insight.

Teng et al.[19] focused on internships and recruitment in the public health space in Singapore in 2021 and found that while recruitments reduced, there were many positives in e-internships. The interesting insight was in the fact that e-interns were able to gain on technical skills and less on soft skills which was understandable.

Adeleke and Gao[1] focus on the changes in the recruitment during COVID-19 period and the lessons learned.

Gupta et al.[7] focus on the recruitment angle of COVID-19 and postulate that aspects such as virtual screening and interviews are here to stay even after the pandemic. This is in line with what we are seeing from the industry.

Similar sentiments are expressed in Vijayan et al.[20] where the opinions of hiring managers were collected. The positives of recruitment during pandemic are increased opportunity, time savings, and the efficiencies. The negatives are the lack of personalized process and need for additional hands-on training for those who are being recruited.

The issues of getting guidance from mentors during COVID-19 were studied by Speer et al.[18] and the experiences showed some interesting results. The communication was expressed to be good but the productivity was by and large worse than their face-to-face experiences.

Hardin and Brooks[8] came up with recommendations for mentoring virtually in this era that stresses clear goal setting, creating anchor events for shared experiences, shape, and build the mentees careers through mock interviews and resume check sessions, etc. These are great suggestions for engagement in this online era.

Qadir and Al-Fuqaha[12] emphasize the principles that a student can follow to succeed and stress goal setting, developing thinking skills, assimilation/getting feedback, taking ownership of learning and development of learning skills. Our work provides the experimental results for the self-motivation aspects and shows that the students who are self-motivated are the ones who have learned most during the COVID-19 pandemic.

The study by Ayadat et al.[3] focused among other things the correlation between teachers' helpful nature and the impressions of the students. Students who thought their teachers were helpful had better feelings about their virtual learning experiences

The impact of COVID-19 on student wellbeing and the coping strategies was studied by Holden and Brown.[9] One third of the participants admitted to being affected by the crisis. Interestingly exercising and social media emerged as the primary coping mechanisms adopted by students.

These above prompted us to identify the primary factor among technology, self-motivation, teaching, family support, and gamified approaches for the students behind their interest and motivation toward online placement learning during the COVID-19 lockdown. We further studied whether lean media methods such as reminders and mails are helpful in this process.

12.3 HYPOTHESES

This section presents the hypotheses for the study

- **H1**: Self-motivation contributes positively to self-efficacy in online distance education in a crisis situation.
- **H2**: Lean media methods contribute to learning.

12.3.1 PRE-COVID-19 SCENARIO

Keeping in mind the recruitment process for the students of the batch of 2017, the placement team of Vellore Institute of Technology (VIT) Bhopal, in association with VIT Vellore planned the training process a few months before the lockdown was implemented. The total number of students preparing for placement from VIT Bhopal was 314.

12.3.2 POST-CLOSURE IMPACT

A virtual platform for conducting real classroom activities was adapted with the aid of online video calling support and moderate network connectivity. During the initial phase, uncertainty started prevailing as students were missing the physical presence of the trainers, one-to-one interaction inside the class, and group activities.

To address the gap in studies, the soft versions of the study materials were being prepared by the teachers and made available online. Course providers such as COURSERA, edX, and Udemy started offering free courses to students. To provide the general idea, scope, and opportunities regarding job sectors, daily webinars were arranged at timed intervals for the students. Online tests were offered daily to sharpen the students' skills and help them understand where they stand.

12.3.3 AIM OF OUR STUDY

As COVID-19 completely changed the style of teaching and training of the students, it became necessary to study the influence of this lockdown on the present way of teaching through the online platform. Placement training plays a vital role in assisting the students to get a job with ease. However, success is achieved not only by the effort of trainers and a well-thought-out plan by the placement teams, but also by the effort and motivation from the students' ends. This study aimed to check whether the students, who were preparing for campus placement, were grabbing the opportunities to pursue free courses offered and understand whether they were facing challenges in performing online tests. We took a survey after each placement training session to help find the answer to these questions. Apart from the quantitative survey, we wanted to correlate the overall data with the present psychological conditions of the students, as it plays an important role in deciding their approach toward the judicial utilization of the newly introduced virtual learning platform.

12.4 METHOD OF SURVEY

The survey and data collections were performed using "Google Forms. " The students were asked periodically via mail and surveys (web forms) to share their views.

The data from the periodic surveys were used to showcase the results. The training started in March 25 and extended till May 17, for which we conducted a series of training events and collected data. On April 4, the first survey was conducted. This established the baseline and we found the information regarding the students'

- demographics,
- connectivity,
- motivation,

- utility of tests,
- their domain study, and
- the reasoning for selecting the particular domain.

On April 17, the second survey was conducted. We found the students'

- strategy of attending the exams,
- state of mind,
- reason for participation in tests, and
- learning outcome and reasoning.

The surveys were conducted again on May 1 and May 17; thus, an overall frame of mind of the learners could be gathered.

12.5 OUTCOMES

The demographics and web connectivity of students, appearing for placement training, are shown below in Figure 12.1A. This shows the distribution of the students in urban, rural, and semi-urban areas, and the challenges they faced. The survey showed that a good portion (66.7%) of the total number of students, resides in urban communities. Interestingly, the web connectivity (Fig. 12.1B) does not directly correlate with the areas. Around 55% of students experience very good connectivity, 25% moderate, and 20% very poor.

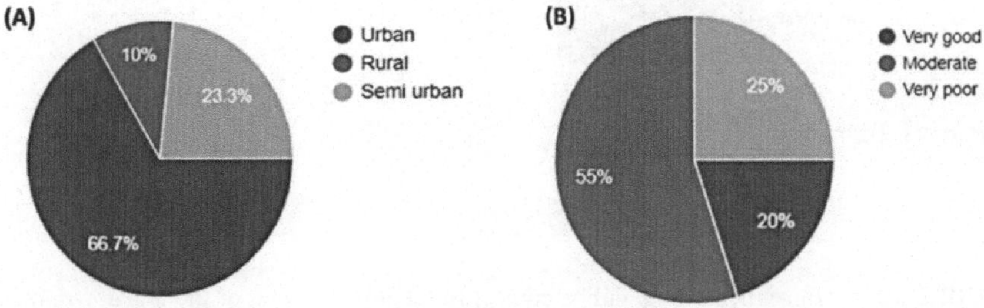

FIGURE 12.1 Location and web connectivity. (A) The distribution of the students in terms of their location. This is a self-categorization by the students. They characterized themselves into urban, rural, and semi-urban localities. (B) The web connectivity of the students. An interesting dynamic here is that being in an urban locality is not an indicator of web connectivity. A large proportion of students reported moderate-to-good web connectivity.

The students were asked whether they had enrolled in free or paid courses (Fig. 12.2A), 15 days after the declaration of lockdown. From the survey, we observed that students in domains having web access have enrolled in 2–4 free courses (Fig. 12.2B). The nature of competitive placement exam preparation is such that the students had an opportunity to pursue courses and they utilized it. The student distribution on our campus is diverse with an equal number of computer science engineering (CSE) and non-CSE branches. The target students were mainly those who are preparing for campus placement and hence courses such as Programming in Data Structures, Java, and Aptitude skills were commonly conducted.

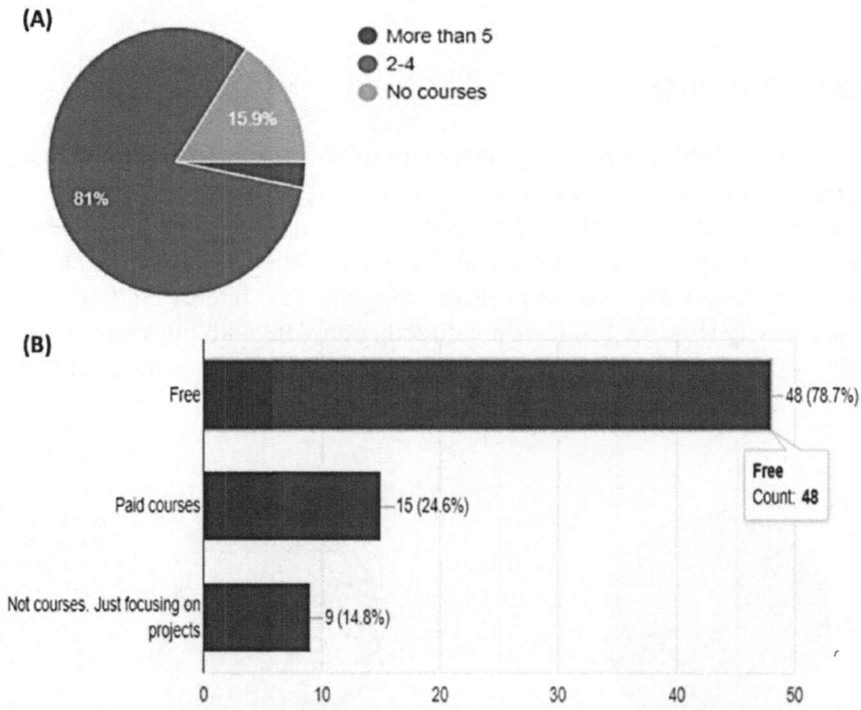

FIGURE 12.2 Enrolment in the online course. In the initial phases of the lockdown, many organizations offered courses free of cost. Students started enrolling in these courses often. The course enrollment trends were seen in (A) with a majority of students enrolling in —two to four courses. We were interested in seeing students whether enrolled just because it was free. (B) We found that a good portion of students enrolled because it was useful for their projects.

The statistics about the completion of the courses and the motivation of students are given below (Fig. 12.3A). Many students were motivated by this chance to learn and completed 2–4 courses. The incompletion rate of students was very low. 16% did not enrol in any courses. So, only 20% of the remaining 80% of students who enrolled in courses did not learn. They completed the courses or used parts of various courses to learn. The primary factor for enrolment was the motivation to learn (Fig. 12.3B). A very delightful aspect that was observed was that the students seemed to be self-motivated to learn, only a few believing that cost was a factor. This augurs well for the course providers as well. This in turn supports our first hypothesis.

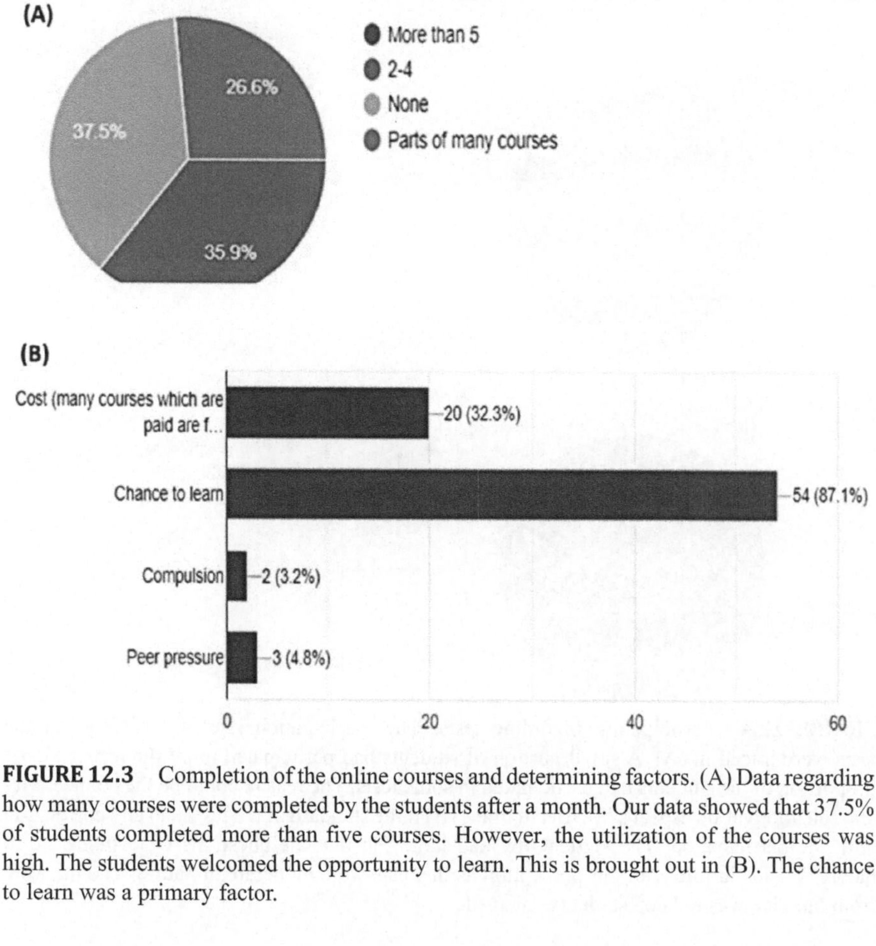

FIGURE 12.3 Completion of the online courses and determining factors. (A) Data regarding how many courses were completed by the students after a month. Our data showed that 37.5% of students completed more than five courses. However, the utilization of the courses was high. The students welcomed the opportunity to learn. This is brought out in (B). The chance to learn was a primary factor.

Various tests were offered to continuously keep a monitor and check up on the pace at which students were learning and understanding. This was because the students were preparing mainly to attend competitive exams of companies. During prelockdown era, we had five tests a week and three of them were proctored with students having a chance to clear their doubts in face-to-face classes. The participation metrics and challenge level of tests are summarized. The students had 16 tests (every day). The majority of the students (40.5% + 32.9%) had attempted a minimum of four tests every day or on alternative days. Strategically the students had focused on targeted learning. The main intent of this study is to analyze and present participation metrics.

An interesting factor we wished to explore was the students' perception of the nature of exams, meaning whether it is very challenging, moderate, or easy (Fig. 12.4B), and their correlation with the participation (Fig. 12.4A).

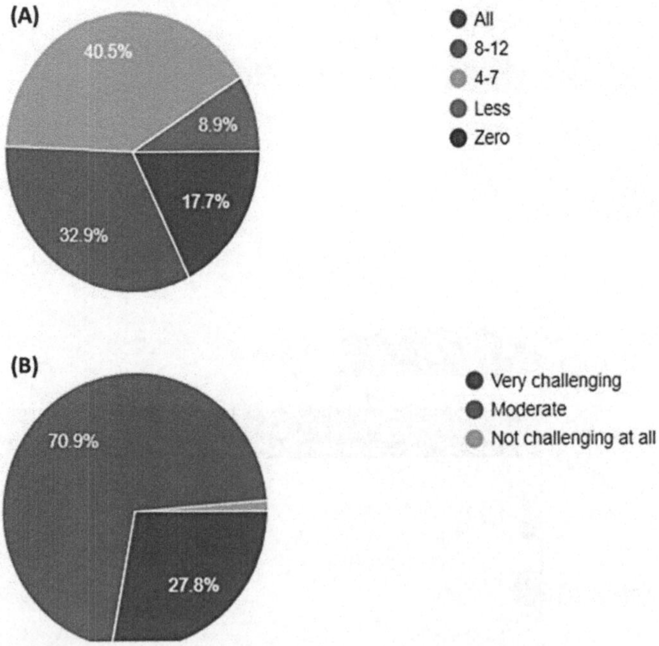

FIGURE 12.4 Participation in online tests. The participation rates by students in the tests were traced in (A). A small portion of students had participated in all the tests. A large proportion of the students had participated in some tests. The reason could be the connectivity and self-motivation aspects. We tried to trace (B) if the standard of the tests had any correlation with the participation. The students by and large felt that the questions were moderate in nature. This is in line with our perception as the tests were moderate in nature. The thinking from our side was to keep students engaged.

We observed that the students seemed to be focusing on the utility (Fig. 12.5A) and self-motivation (Fig. 12.5B) aspects rather than the challenging aspects of attempting those online exams (Fig. 12.5A). The utility of tests and motivation are summarized in Figs. 12.5A and 12.5B. The students were seemed to be motivated to participate in the exams regularly and were not swayed by other factors supporting our hypothesis H1.

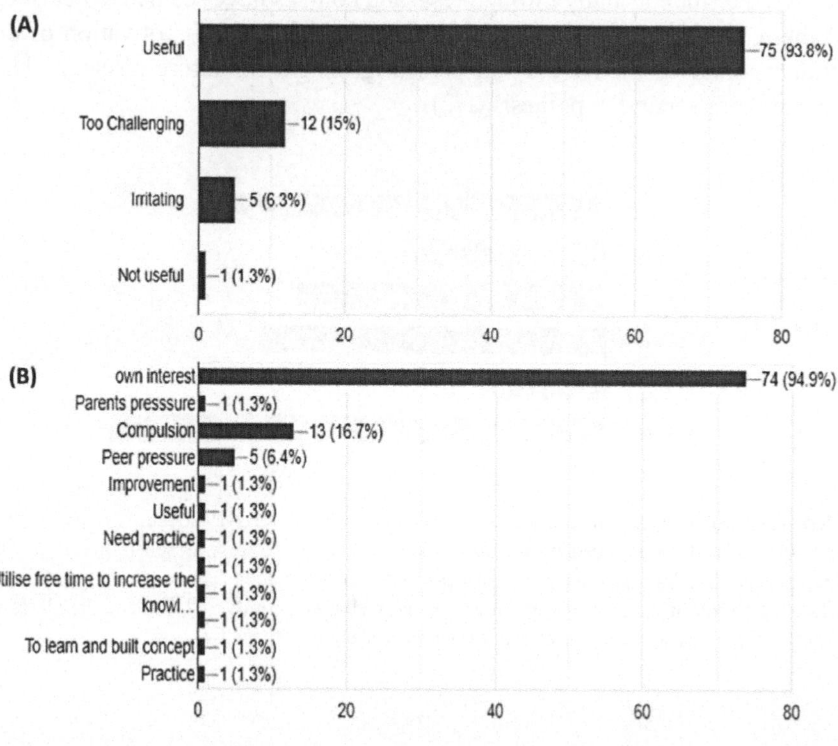

FIGURE 12.5 Learning through online platforms—a self-driven process. (A) The data represents the survey that how the students think about the online training methods. Data showed that students find it useful for their knowledge enhancement for placement training. (B) The attempt was made to find out the main driving factor behind the participation and completion of online training tests. It was observed that the whole process is mainly driven by the students' interests.

The above survey suggests that in learning, the biggest motivational factor is the student's interest. This proved our first hypothesis that student motivation is the primary factor. This was supported by the data given in Fig. 12.6. Additionally, we found that apart from self-motivation, lean media such as daily mail and external stimuli play important roles in pushing the students toward online learning (Figs. 12.6 and 12.7). The above and below bar diagrams (Figs. 12.6 and 12.7) which focus on email reminders, showed that lean media strategies can affect online learning. One other visible aspect we observed in many tests was that these lean media strategies of reminders make a difference in ensuring that students who are not motivated, also attempt the exams (Fig. 12.7). It was also observed that the strategies to engage the students daily through tests and courses have seemed to be useful to them, which proves the driving factor behind the self-motivation of the students (Fig. 12.8). This is an interesting insight for future research. This supports our second hypothesis (H2).

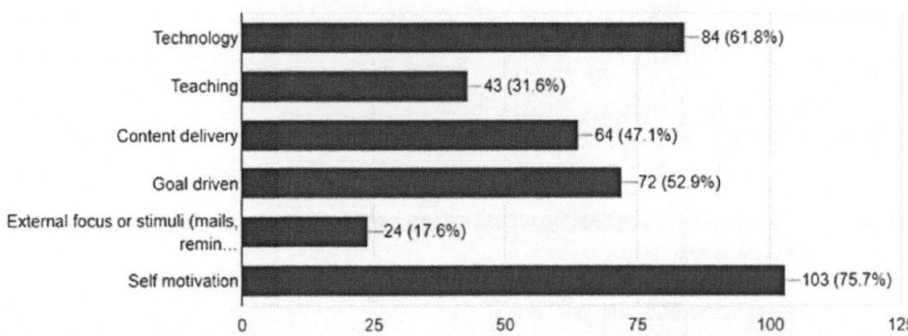

FIGURE 12.6 The essential aspects of learning amid COVID-19 lockdown. We wanted to trace the challenges faced by the students. The biggest factor was self-motivation. The students felt that the best strategy for success was to be motivated. The second biggest factor was the stability of the technology. This included web connectivity-stability and the platform stability. Teaching and content delivery were identified after these two.

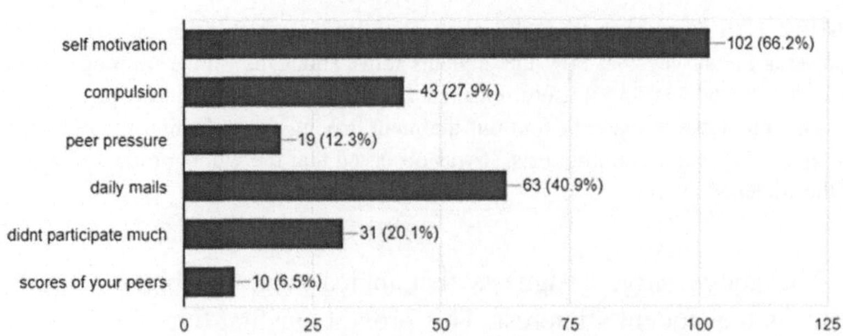

FIGURE 12.7 The reason behind attending online quizzes and vendor courses.

Courses were offered by vendors appointed by the university. We wanted to confirm whether students were participating due to their interest or due to pressure. While we were motivating the students to attend through regular emails, the students felt that their motivation was a big factor. The stimuli through mail also seemed to play a role. The emails gave a reminder and then the students could decide if the session was worth it or not. This is an interesting insight as from a student perspective they seemed to say that learning in this environment needed a strong self-motivation and belief from themselves. The results support our primary hypotheses H1 and H2.

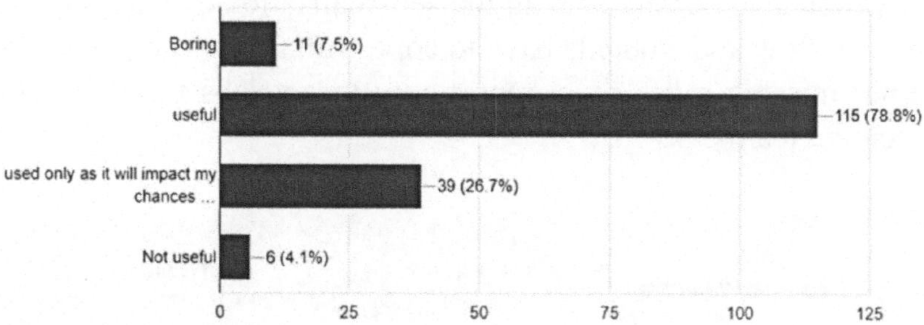

FIGURE 12.8 The strategies of daily engagement through online tests.

We wanted to test if the daily process was useful or not. The students seemed to think in line with their motivation that a sustained effort was useful for their success.

Further, we wanted to find out a strategy that would be preferred by the students during the lockdown period, going in parallel with their "self-motivation." Our data survey showed that self-learning, based on the courses offered by vendors and other free providers, is mostly preferred by students (Fig. 12.9). Apart from that, test-based studies and brushing up on learning also play a determining role in this regard.

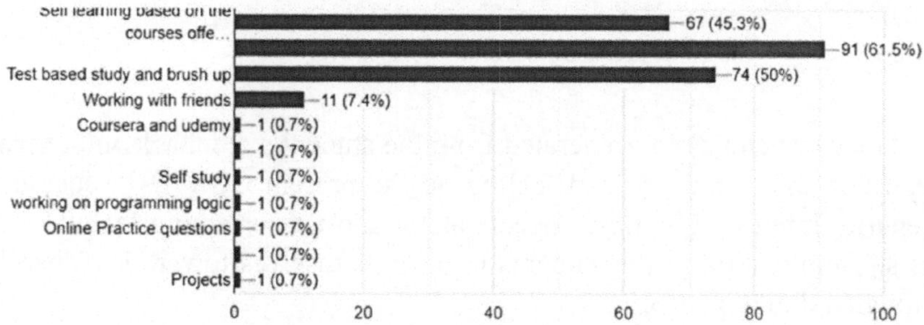

FIGURE 12.9 The preferred strategies to learn in this crisis period.

We wanted to test the impact of peers. It was observed that in COVID-19 scenario the courses offered by vendors and other providers at free of cost were the primary strategies for learning. In line with the focus of the students, the emphasis was on the test-based study and brushing up. The scope for peer study was limited.

Apart from quantitative analysis, the word cloud was generated from students' feedback and thoughts (Fig. 12.10) on how this period has changed their lives. Interestingly, it mainly focuses on time, which is the ultimate determining factor in our life. We believe that self-motivation also correlates with the self-realization of individuals that time will flow and students have to cope with time to ensure that they are not missing a life-time opportunity to get jobs through campus placement teams.

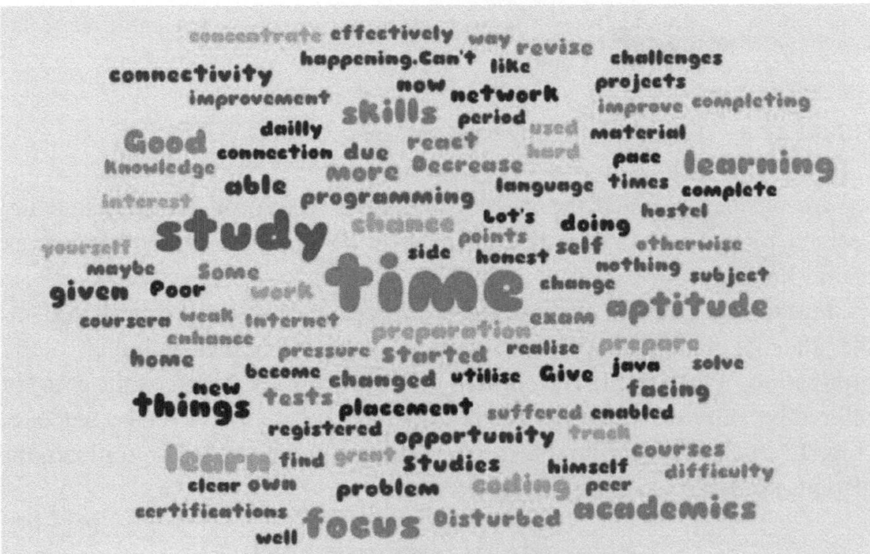

FIGURE 12.10 Cloud of thoughts – student's perspective.

The word cloud was generated from the student's feedback and remarks regarding their thoughts and feeling on the present COVID-19 lockdown scenario. Interestingly, time appears as the most determining factor behind the self-motivation of the students to push themselves toward self-learning through a digital platform.

12.6 DISCUSSION

Participation in the survey was completely voluntary. Out of 314 students, 60% of students participated in our survey. These were the same groups of students who had participated regularly in all the placement activities in pre-COVID-19 time. Hence the opinions were inherently from the regular participants of the activities. This study is an attempt to understand the motivation and challenges faced by a group of students when their learning routine was unexpectedly transformed into the virtual one. Here, we have highlighted the facts faced by students, which came out from our survey report:

- A large number of students enrolled in the courses
- On their motivation (as opposed to any compulsion or pressure)
- Participation in academic aspects like tests and online learning

The challenges faced by the students are in

- completion of learning;
- motivation to study in a time of uncertainty (from the word cloud);
- worries about post-lockdown future.

Main complications faced by students:

- Technology
- Delivery of content
- Teaching
- Indirect compulsion to get involved

Factors appreciated by students:

- Focused content
- Delivery
- Teacher's efforts
- Lean media approaches like mail

12.7 LIMITATIONS

The work was done in our university over 1 batch of students over a period of 3 months. This is a big limitation. Ideally, we must have done this over 2 batches of students and may be a few more institutions as well. However, the training method of our university was different and focused on lean media and gamified approaches. So, the broader aspects could not be covered. Also, we focused on a period of 3 months only. This was because our efforts focused on training alone. In future, the same study can be planned with larger number of students.

12.8 CONCLUSIONS

Students expressed that during COVID-19 lockdown, self-motivation, goal-driven nature, content, and technology were essential and crucial aspects that judge whether the online training platform would be successful or not. Our study represents the issues and challenges faced by the students appearing for placement training during COVID-19 lockdown. We believe that our study will provide important insight into both, the academic and student communities, regarding the challenges faced by students during placement training and provide them with useful strategies to boost their self-motivation.

ACKNOWLEDGMENT

SR and BB are especially thankful to VIT Bhopal University and Adamas University, respectively, for their continuous support in this work.

CONFLICT OF INTEREST

The authors declare that they have no conflict of interest. All the co-authors have given their consent for this work.

KEYWORDS

- **online learning**
- **placement training**
- **self-motivation**

REFERENCES

1. Adeleke, S.; Gao, C. COVID-19 and its Impact on the Clinical Specialty Training Recruitment Process: Lessons Learned and the Shape of Future Specialty Recruitment in the UK. *J. Royal Soc. Med.* **2021,** *114* (6), 323–326.
2. Agarwal, S.; Kaushik, J. S. Student's Perception of Online Learning During COVID Pandemic. *Indian J. Pediatr.* **2020,** *87* (7), 554–554.
3. Ayadat, T.; Khasawneh, M. A.; Chowdhury, S. R.; Nayeemuddin, M.; Ahmed, D.; Asiz, A. Effects of Students' Home Environment, Tools, and Technology Used on Online Learning Experience in a Civil Engineering Program. *Int. J. Inf. Educ. Technol.* **2021,** *11* (8).
4. Bal, I. A.; Arslan, O.; Budhrani, K.; Mao, Z.; Novak, K.; Muljana, P. S. The Balance of Roles: Graduate Student Perspectives During the COVID-19 Pandemic. TechTrends **2020,** *64* (6), 796–798.
5. Compton, S.; Sarraf-Yazdi, S.; Rustandy, F.; Radha Krishna, L. K. Medical Students' Preference for Returning to the Clinical Setting During the COVID-19 Pandemic. Med. Educ. **2020,** *54* (10), 943–950.
6. Fansury, A. H.; Januarty, R.; Rahman, A. W.; Syawal, S. Digital Content for Millennial Generations: Teaching the English Foreign Language Learner on COVID-19 Pandemic. *J. Southwest Jiaotong Univ.* **2020,** 55 (3).
7. Gupta, S.; Arthur, L. G.; Chandler, N.; Danielson, P.; Downard, C.; Ehrlich, P.; ...; Mak, G. Is the Changing Landscape of Fellowship Recruitment During COVID-19 Here to Stay?. *J. Pediatr. Surg.* **2021,** *57,* 445–450.
8. Hardin, A.; Brooks, S. In *Pivoting to Virtual: Recommendations for a Successful and Engaging Digital Mentorship Experience*, The 39th ACM International Conference on Design of Communication, 2021; pp 351–353.
9. Holden, C.; Brown, T. Students in Graduate Program Share Coping Strategies Utilized during the COVID-19 Pandemic. *Open J. Soc. Sci.* **2021,** *9* (7), 15–21.
10. Holt-White, E.; Montacute, R. COVID-19 and Social Mobility Impact Brief# 5. Graduate Recruitment and Access to the Workplace, 2020.
11. Izumi, T.; Sukhwani, V.; Surjan, A.; Shaw, R. Managing and Responding to Pandemics in Higher Educational Institutions: Initial Learning from COVID-19. *Int. J. Disaster Resil. Built Environ.* **2020,** *12* (1).
12. Qadir, J.; Al-Fuqaha, A. A Student Primer on How to Thrive in Engineering Education During and Beyond COVID-19. *Educ. Sci.* **2020,** *10* (9), 236.
13. Kaur, D.; Sahdev, S. L.; Chaturvedi, V.; Rajawat, D. Fighting COVID-19 With Technology and Innovation, Evolving and Advancing With Technological Possibilities. *Int. J. Adv. Res. Eng. Technol.* **2020,** *11* (7), 395–405.
14. Kock, N.; Davison, R. Can Lean Media Support Knowledge Sharing? Investigating a Hidden Advantage of Process Improvement. *IEEE Trans. Eng. Manag.* **2003,** *50* (2), 151–163.
15. Kvan, T.; Gao, S. *Frames, Knowledge and Media - An investigative Study of Frame Systems Within Computer and paper Supported Collaborative Design process,* In Architecture in the Network Society, 22nd eCAADe Conference Proceedings, Copenhagen, 2004; pp 410–417.
16. Lall, S.; Singh, N. COVID-19: Unmasking the New Face of Education. *Int. J. Res. Pharm. Sci.* **2020,** 48–53.

17. Oregon, E.; McCoy, L.; Carmon-Johnson, L. Case Analysis: Exploring the Application of Using Rich Media Technologies and Social Presence to Decrease Attrition in an Online Graduate Program. *J. Educ. Online* **2018,** *15* (2), n2.
18. Speer, J. E.; Lyon, M.; Johnson, J. Gains and Losses in Virtual Mentorship: A Descriptive Case Study of Undergraduate Mentees and Graduate Mentors in Stem Research During the Covid-19 Pandemic. *CBE—Life Sci. Educ.* **2021,** *20* (2), ar14.
19. Teng, C. W. C.; Lim, R. B. T.; Chow, D. W. S.; Narayanasamy, S.; Liow, C. H.; Lee, J. J. M. Internships Before and During COVID-19: Experiences and Perceptions of Undergraduate Interns and Supervisors. Higher Education, Skills and Work-based Learning, 2021.
20. Vijayan, J.; Siby, D. A.; Sabeen, G. P. In *Impact of Covid-19 Pandemic on Recruitment Process and its Sentiment Analysis,* 2021 2nd International Conference on Advances in Computing, Communication, Embedded and Secure Systems (ACCESS); IEEE, 2021; pp 270–273.
21. Verma, G.; Priyamvada, M. COVID-19 and Teaching: Perception of School Teachers on Usage of Online Teaching Tools. *Mukt Shabd J.* **2020,** *9,* 2492–2503.
22. Woywodt, A.; Breed, H.; Lumsden, C. COVID-19–The Ultimate Disruptor? *MedEdPublish* **2020,** 9.

CHAPTER 13

Implementation of a Postpandemic Crowd Safety Provisioning Application Using AI Methods

ALANKRITA AGGARWAL and SHIVANI GABA

Panipat Institute of Engineering and Technology, Samalkha, Haryana

ABSTRACT

The universe has gone through the worst pandemics, and even now, the situation is unclear, and people face many consequences. No one has thought that this kind of pandemic would come into our lives. These pandemics have ruined many lives. Even after multiple lockdowns, more people are heading toward the streets and are on the verge of risking their lives. The pandemic has led to a harmful impact on the generation. This work proposes an innovative approach toward providing safety to the end user, that is, the general public. The authors have implemented a postpandemic crowd safety provisioning application using artificial methods. As we know we cannot disrupt the economy; we have to get back on track, and the virus is nowhere at its termination. So, the proposed system notifies the user about all possible harmful locations and further provides a safer route to travel via so that the user can safely or at least be at the minimum risk of contracting a virus-prone zone during their journey.

13.1 INTRODUCTION

The world has gone through and still facing one of the worst pandemics. After multiple lockdowns, more people are heading toward the streets and are on the verge of risking their lives; hence the proposed system notifies the

Global Digital Transformation and the Covid-19 Pandemic.
Rajdeep Chowdhury, K. Martin Sagayam, Neyara Radwan, Ahmad A. Elngar (Eds.)

user about all possible harmful locations and further provides a safer route to travel via so that the user can safely or at least be at the minimum risk of contracting a virus prone zone during their journey. The user will open the app on whichever device they prefer. The app will use the current location coordinates of the user to determine where the corona-prone zone is and alert the user in turn. The primary purpose of the work is to target the users for whom any system is developed with the thought of an end user in mind. This proposed work explained in this chapter, is globally usable by any person irrespective of the field; any person who has to get to someplace and is keen to know the current COVID-19 situation around their location is an ample user for this product being developed.[1–3]

We already know the Internet is the mother lord of the modern world. Everyone has been interlinked for the last two decades or so. Earlier, we used to use computers and other devices, but after the mobile market boom, everything has become easy to use and is within reach of our fingertips. With functionalities like maps, anyone can access facilities like knowing about a location, going to a particular place, rating and reviewing, and much more. However, in the past few months, a severe pandemic has hit the world, which has curbed everyone to their homes. After rigorous research and analysis, we have deduced that the pandemic is nowhere near its termination. Hence keeping the current scenario in mind, people, after all, have to get to the streets for fulfilling their basic needs such as buying groceries, and a necessary task such as going to your firm. It also includes delivery personnel, who are quite popular these days, so, keeping in mind all these situations in our mind people would love to get their jobs done in the safest ways possible when they hit the road, and it would be more convenient if a device that is always in their hands could be used for precisely this.[4–8]

13.1.1 MOTIVATION

In the current pandemic situation, even after rigorous research and hard work, we have not found a permanent end to this COVID-19 virus, and the safety of every individual is at stake. Also, we cannot let everyone sit at home; although most of the tasks have shifted to work from home, the economy has shown us what destructive impact the current situation has laid on us.

- People have to get to the streets to fulfill their basic needs, such as buying groceries for necessary tasks such as going to your firm.
- It also includes any delivery personnel, which is quite popular these days, so keeping in mind all these situations in our mind people

would love to get their jobs done in the safest ways possible when they hit the road, and it would be more convenient if a device that is always in their hands could be used for precisely like this.[9–12]

13.1.2 *CONTRIBUTION*

In the past few months, a severe pandemic has hit the world, which has curbed everyone to their homes, and even after rigorous research and hard work, we have deduced that the pandemic is nowhere near its termination. So, we have:

- The "postpandemic crowd safety" app enables its users to know about the current COVID-19 situation,[40] that is, how many hotspots are there in the surrounding region, whether a place safe enough to visit or not and what could be safest path to go from a location "A" to site "B," where could be a potential hotspot due to overcrowding of a zone.
- All these are just at ease of access of your fingertips. However, the system is not only limited to that.
- With this system being developed, a user can access all the information mentioned above from multiple platforms such as Android, IoS, and Web using a simple application or simple Web app.
- The application has a minimal interface so that anyone with basic mobile operating knowledge can utilize the system to understand the current pandemic situation.[13–19]

13.2 RELATED WORKS: RATIONALE BEHIND THE NEED FOR THE SYSTEM

No one ever thought about how this world would happen before March 2020, and our whole life has been disrupted. Everyone is concerned about physical and emotional energy, employment, money, and health.

As we cannot disrupt the economy further, everything has to get back on track, and the virus is nowhere terminated; hence, we need to find an excellent method to ensure the general public's safety.[20–29] The potential benefits derived upon implementation, the system could bring about significant tangible and intangible benefits:

TABLE 3.1 Tangible and Intangible Benefits.

Tangible benefits	Description	Intangible benefits	Description	Reference
No investment	Users, do not have to pay anything and need an Android mobile platform and are flexible to use any interface.	Reduces effort	It reduces the effort, for example, to bring all the surrounding hotspots in real time.	[30]
Ease of access	This system can be accessed anywhere; whether it is Android, IoS, or also through Web app.	Time saving	The user is fed with the appropriate information as soon as the app launches, he/she does not have to fiddle around with the options to get information.	[31]
Easy to use interface	The application will be minimalistic yet intuitive to use so that even a layman can use it with ease and inefficient way.	Friendly user interface	The application that is developed would facilitate its end users with interactive graphical interfaces that would be easily adaptable.	[32]
Accurate Geolocation	Upon permission agreement, user is fed with accurate Geolocation of himself as well as surrounding hotspots.	Transparency	This system provides transparency in hotspot determination. Every hotspot or crowded zone determined is shown irrespective of the place.	[34]

13.3 OBJECTIVES OF THE SYSTEM

The current system aims to provide safety to the general public using digitization and ease of access so that an end user can quickly determine any crowded zones or corona hotspots from their device itself. To achieve this team has devised the following objectives:

- Since this application is cross-platform and keeping in mind most of the users, we have prioritized the development of an Android application, an IoS application, and then a progressive Web app.
- Since the user needs to determine a nearby hotspot, we must implement geolocations services.
- The system also needs to notify users regarding any crowded zones.
- There is also an utter need to provide the safest path between a source and a destination.
- Assume that every 8 out of 10 people have a smart device on which the application program will execute.

The system allows the developer to acquire information specializing in cross-platform software development and automation. The developer underwent software development, proposed work, explained in this chapter management, human–computer interaction (HCI), usability factors, and the mechanisms of development principles to produce highly reusable quality software on time.[38,39] However, the prime aim of undergoing mobile computing concepts is to increase preferred output with user satisfaction within less effort and time to encounter an easy solution for checking corona hotspots nearby them and reduce all the steps required for knowing the current details of a hotspot zone. The following objectives are of prime importance to the developer:

- They are gaining knowledge of Web-based development, Web technology, and JavaScript Object Notation (JSON).
- Learning the concepts of services such as Web services and location-based services.
- Learning how to deal with web-based software and development and its database management.
- Able to perceive real-time applications with their integration on a computer/laptop.
- Implementation and integration of database services with Web applications.
- Understanding BLoC—for a robust architecture.
- Understanding data gathering methods and indexing it into the database of relevant style.
- A better understanding of proposed work is explained in this chapter management concept, that is, scheduling, budgeting, cost estimation, work breakdown structure, etc.
- Learning and practices of Docker, Kafka, and PostgreSQL.

13.3.1 PASS/FAIL CRITERIA

- The produced system meets all the specifications mentioned in the proposed work, explained in this chapter specification form, and all the user requirements gathered during the research phase.
- All the modules/functionalities are implemented so that the desired system produced runs without any bugs and quickly.
- The system is to be verified for usability that depends upon proven HCI principles and aims to achieve ease of use, navigation, time for completion, and error removal in the application.

13.3.2 CRITERIA OF TEST CASES

The system must comply with the success criteria mentioned in the specification. Some of the requirements are:

- The product management module should work efficiently.
- Transaction management module should work efficiently.
- Synchronization must work efficiently.
- The system should satisfy all user requirements.
- The Website should be easy to use.

13.4 RESULTS AND DISCUSSIONS

The following are the snapshots of the work which is presented as an app on the mobile phone of every user to detect the crowd and will alert the crowd to scatter. Following are the snapshots of the proposed work, explained in this chapter work carried out while making and testing the app and Web application, and the snapshot of the user interface of our work is shown in Figure 13.1.

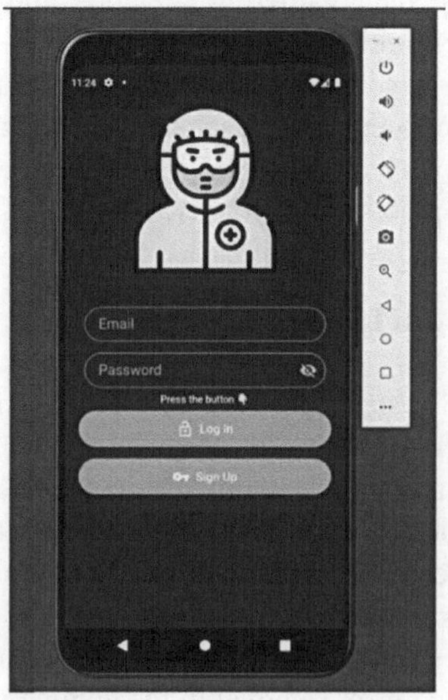

FIGURE 13.1 Snapshot of user interface.

As the application is the implementation of a postpandemic crowd safety provisioning application using AI methods, it is the application has a login view, signup view, error detection field view, and final signup view as shown in Figures 13.2. 13.3, 13.4, and 13.5, respectively.

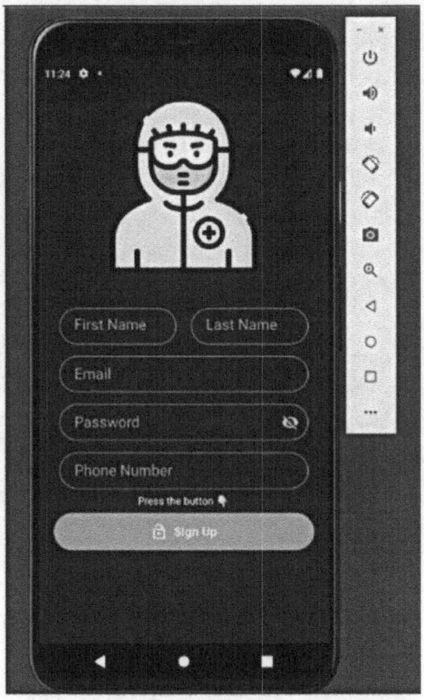

FIGURE 13.2 Login view.

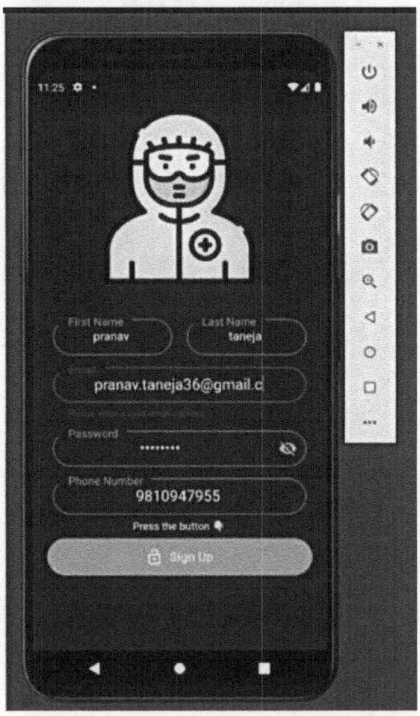

FIGURE 13.3 Signup view.

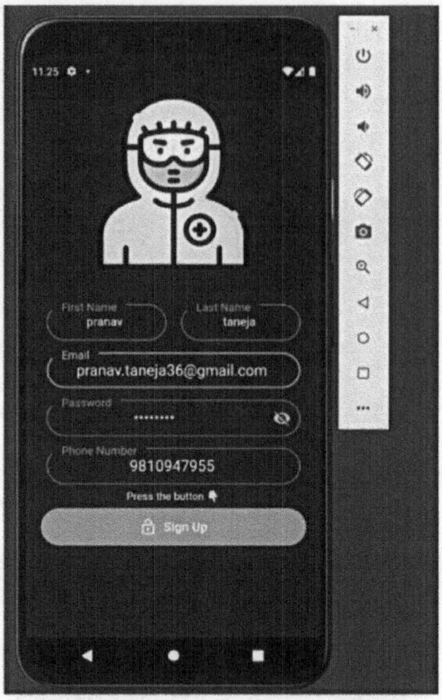

FIGURE 13.4 Error detection on fields.

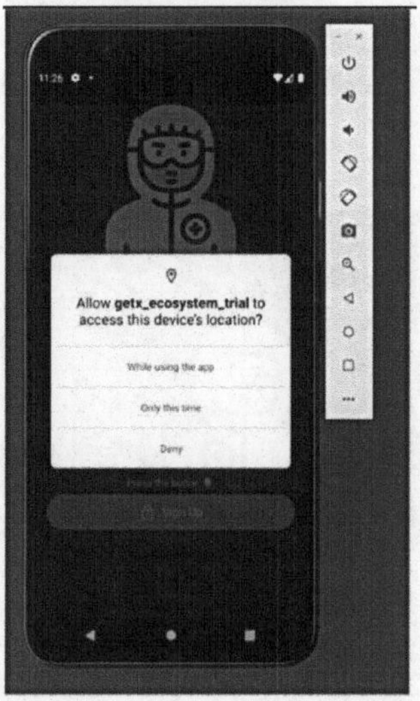

FIGURE 13.5 Final signup view.

Afterward, the GPS permission dialog, the Map View with Hotspot Zones, Details of Selected Hotspot Zone, and the overview of Crowded Zones are shown in Figures 13.6–13.9.

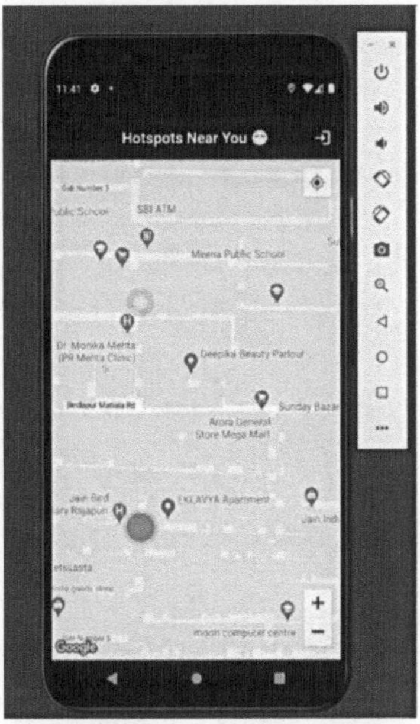

FIGURE 13.6 GPS permission dialog.

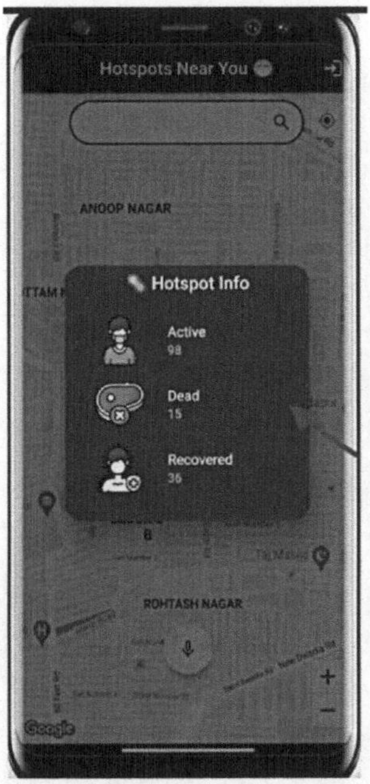

FIGURE 13.7 Map view with hotspot zones.

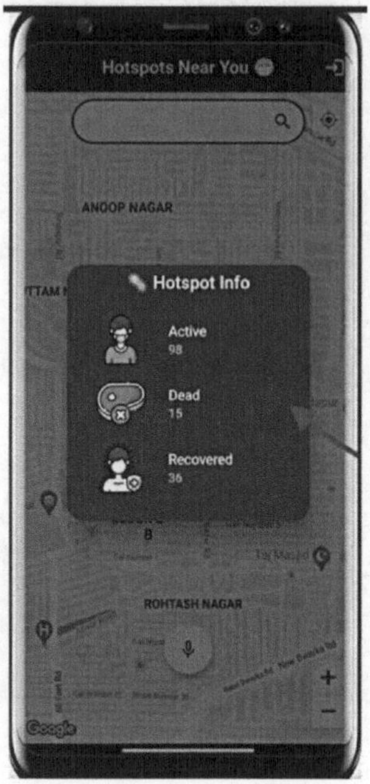

FIGURE 13.8 Details of selected hotspot zone.

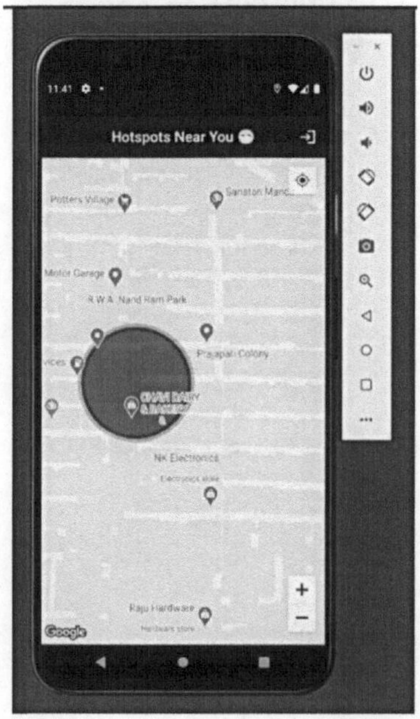

FIGURE 13.9 Crowded zones.

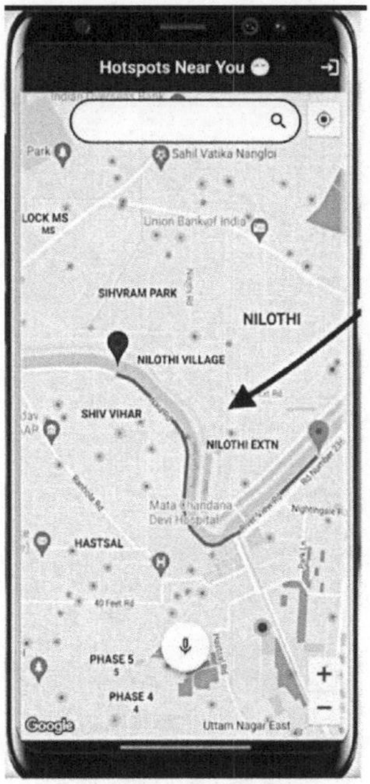

FIGURE 13.10 Safest route.

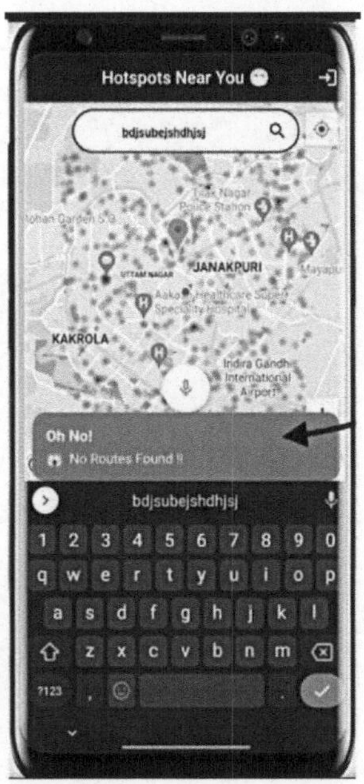

FIGURE 13.11 Error notification.

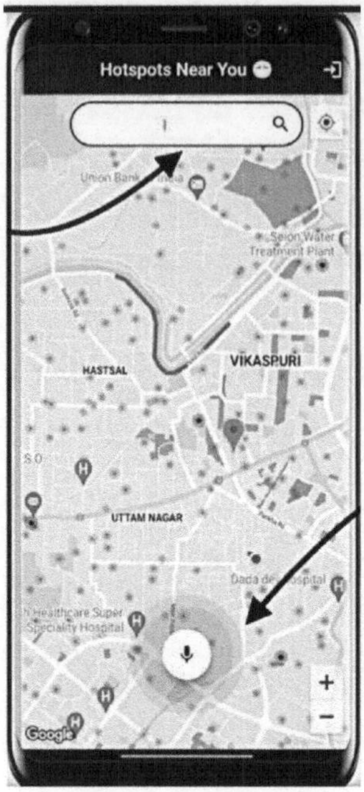

FIGURE 13.12 Voice search.

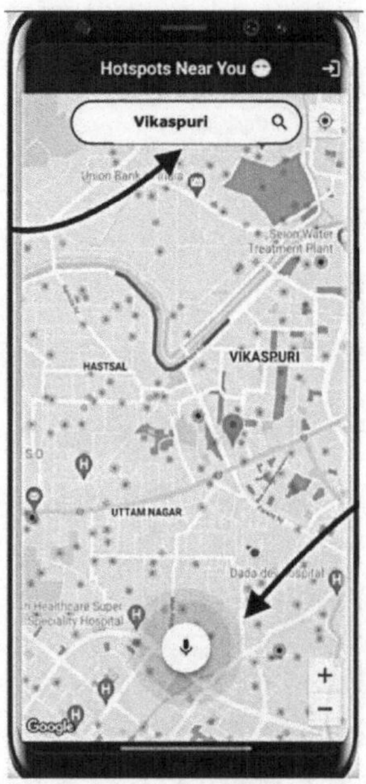

FIGURE 13.13 Custom search.

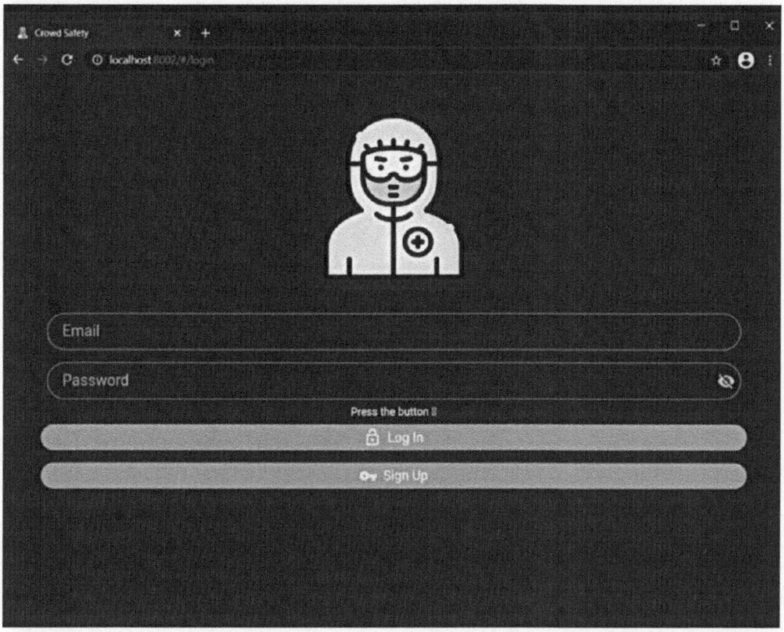

FIGURE 13.14 Login Web view.

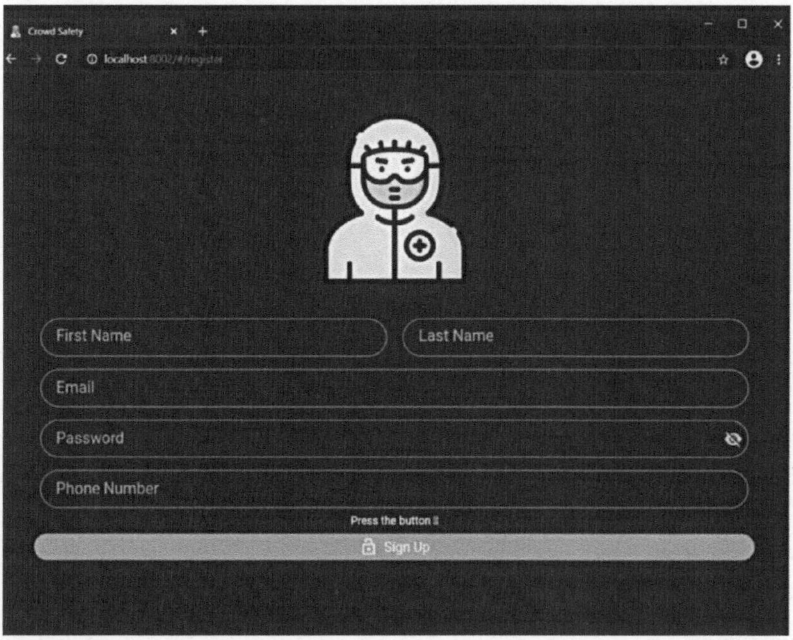

FIGURE 13.15 Signup Web view.

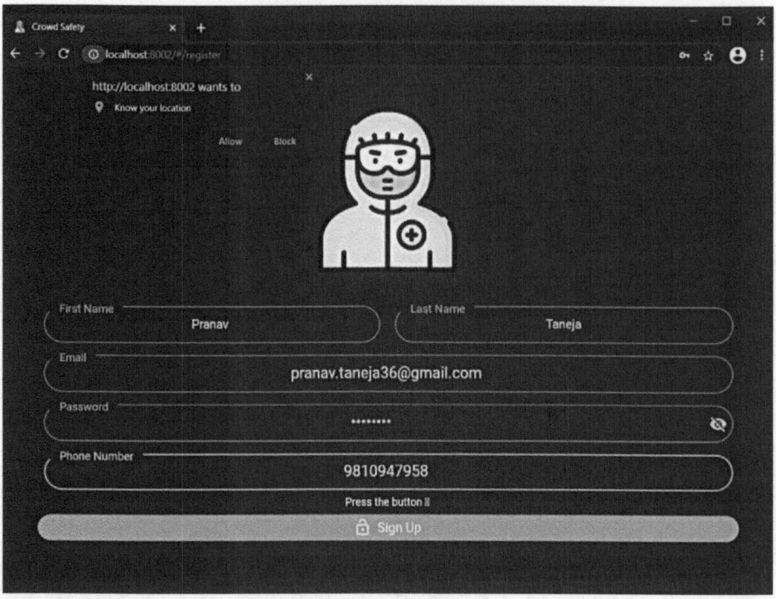

FIGURE 13.16 Permission Web view.

FIGURE 13.17 Hotspot Web view.

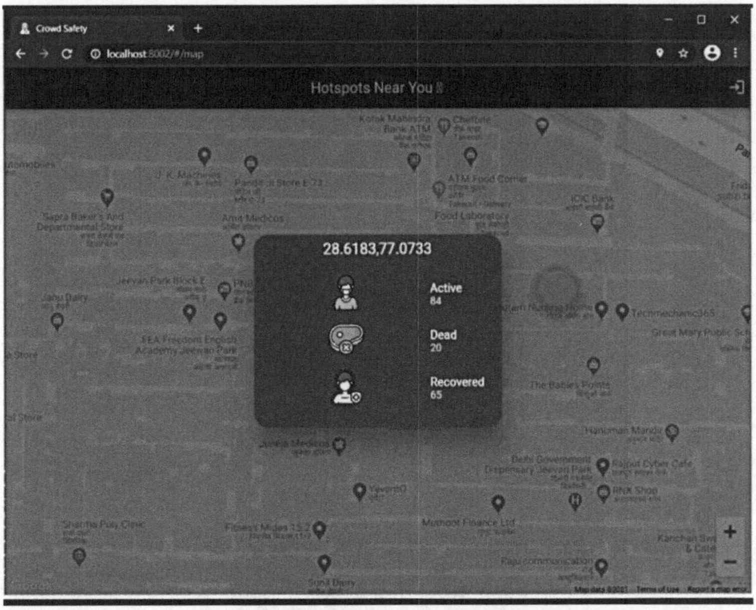

FIGURE 13.18 Hotspot detail Web view.

FIGURE 13.19 Crowded hotspot Web view.

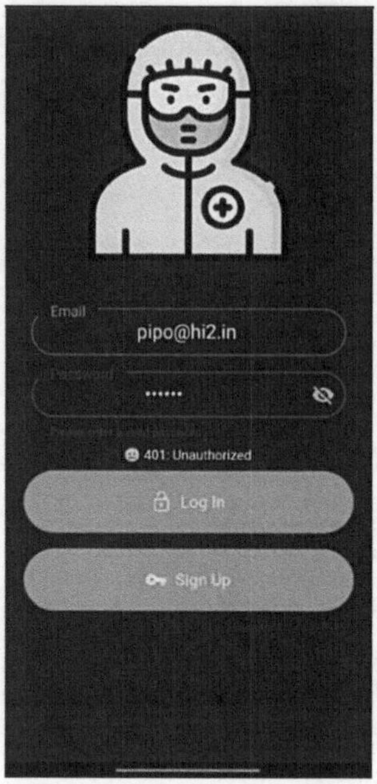

FIGURE 13.20 Authorization.

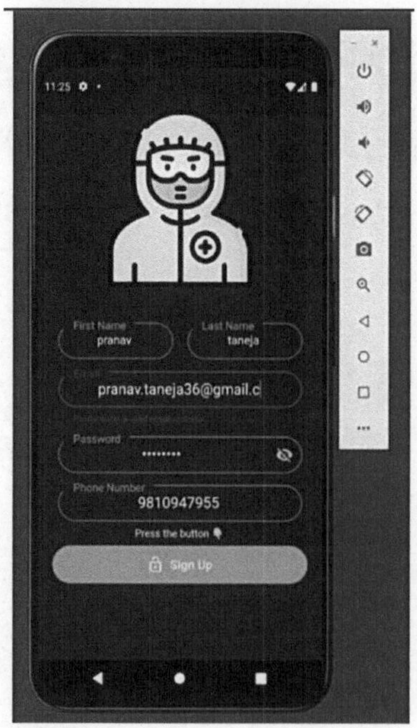

FIGURE 13.21 Field validation.

Moreover, how the user can take the safest route is shown in Figure 13.10 and if the errors come, the notification of errors is shown in Figure 13.11. The voice can also be searched and the snapshot of that is shown in Figure 13.12. If the user wants to do searching in a custom way, that is shown in Figure 13.13.

The snapshot of the login Web view and signup Web view is shown in Figures 13.14 and 13.15, respectively. In continuation to that, the snapshot of permission of web view, a hotspot of web view is shown in Figures 13.16 and 13.17, and the hotspot detail web view and crowded hotspot web view are shown in Figures 13.18 and 13.19, respectively. And finally, the snapshot of the authorization and field validation view is shown in Figures 13.20 and 13.21.

13.5 SUCCESS CRITERIA

The proposed framework called the "postpandemic crowd safety system" is primarily valuable for the end customer as it meets all the details of client requirements just as it satisfies every model that judges quality again the implementation of the framework. The performance of the proposed framework is achieved as it has met the process of success and created the right one. The framework has resolved the issues assessed in Section 2 relating to issues suggested and arrangements provided. The framework, after testing, promises to provide security for its users by giving compliant details about the current epidemic for them.

13.5.1 LIMITATIONS AND ERRORS IN THE DEVELOPED SYSTEM

- The customer will need an Internet connection to get information simultaneously about the current situation.
- The entire proposed work explained in this chapter is based on the crowdsourcing of data.
- The user should have GPS enabled on their device.

13.5.2 LEARNING EXPERIENCE GATHERED

Learned a lot from research work:

- In the first phase of the work, the developer was confused about what had to be remembered for the framework.
- In this way, the designer completed the optional study by considering the experimental activities of various creators and watching to pay attention to all the mistakes they have made.

- It helped find unusual information about the story and comparative distribution structures related to their prominence and decline.
- Finally, it is led by necessary designer research to integrate ongoing customer needs.

13.5.3 LEARNED TO MANAGE STRESS AND TIME

While working at various times of software development lifecycle (SDLC), the engineer has taken on different functions such as the following: analyzer, analyst, database, and portable app designer and analyzer. This gave the designer the power to manage pressure and time successfully. The engineer had a chance to finish the job within the time frame we overcome all the burdens and needs of the opportunities during development.

13.5.4 GAINED TECHNICAL EXPERTISE

- The framework's usage required not exclusively to apply learning results accomplished during scholastics but also a reasonable comprehension of different versatile innovations.
- It required much web searching to grasp the user interface design and the structure of the system. It led to a huge learning curve for the developer.

13.5.5 HAD AN EXPERIENCE OF PROPOSED WORK, EXPLAINED IN THIS CHAPTER MANAGEMENT

- The undertaking gave an excellent encounter and sentiment of filling in as though in some corporate world.
- The designer likewise assumed the imperative job of an undertaking supervisor by getting the finished-on schedule, conveying a quality framework, satisfying client necessities, and utilizing reality-discovering strategies to gather prerequisites.
- The separation and vanquish approach was utilized to partition the whole venture into little assignments and finish them as indicated by the calendar arranged in the Gantt graph.

13.5.6 HOW TO RESOLVE ERRORS AND LEARN FROM THEM

- The designer has encountered many errors and exceptional cases reviewed in the background using different techniques when using an outline.

- It provides a glimpse of how to solve a problem and fix errors using the top–down method.
- Similarly, an engineer saw a way to find errors and correct them, which is appropriate and is expected to pass an error-free outline.

13.5.7 GAINED MUCH CONFIDENCE

Before outlining, the developer was in a particular position. How to use complex framework modules. Be as realistic as possible; continuously, things became more accessible after doing the research work in each development period itself, removing the concept-building approach and spreading the whole framework. Still, it required a steady and honest effort, yet the designer proved to be the idea that "where there is a will, there is a way." This study inspired the composer, who embraces the unique needs by unquestionably accepting that there is an answer to each issue.

Last but not least, this last year's venture gave an extraordinary learning and captivating experience of chipping away at ongoing ventures and made a profound enthusiasm for the portable application advancement field, which will help in receiving reward for having a splendid profession in the future.

The above work shows the proposed framework and how the framework informs various stages as we progress or move. Here the authors began with an essential investigation to determine whether there is a need for a security system to provide users with information about its COVID-19's worst areas. The primary reading phase is followed by a literary study of the type of research to integrate data into the mitigation site and focus flow to build a functional framework. The designer led necessary and helpful research. Assuming it is a good combination of data about the proposed framework, the designer needed to hear a different development.

Surveys, discussions, and data collection methods are used to identify the needs of willing people and workers within the organization. Research is completed by staff/staff, while management and in-depth reception lead interviews. Poll testing, meetings, and institutional circles helped the developer to master its full details and to understand the organization's security efforts as clients were taken from various machine parts. In addition, unique details were handled by meeting with product engineers and planners. Their suggestions check for internal and external information on developments and details. These lines have completed the test portion of the proposed work. The engineer then proceeded with the part of the draft frame where the draft adjusted configurations that include the intellectual and physical systems of the framework that advance on unified modeling language (UML) lists, framework body plans, relationship graphs, and more.

It comes to the point of use, which includes dealing with compliance with a designer to create different modules for a framework that starts with a lower module need and goes toward the highest requirements. Otherwise, the client manual as a particular manual is provided that assists the client with the continuation of the framework and how to use the framework and direction of the establishment separately. The final task is assessed to determine the decision, how the framework can benefit its end clients, and how the work benefits the client's perspective. Finally, the last task is assessed to arrive at a resolution, how the framework can benefit its end clients, and how much the undertaking is fruitful from the client's perspective.

13.6 CONCLUSION

As in the current pandemic situation, even after rigorous research and hard work, we have not found a permanent end to this COVID-19 virus, and the safety of every individual is at stake. Also, we cannot let everyone sit at home; although most of the tasks have shifted to work from home, the economy has shown us what destructive impacts the current situation has laid on us. So, the proposed system will be beneficial in several ways. The "post-pandemic crowd safety" is about safely letting users know about their current surroundings by providing detailed information about the current pandemic situation around them. So, ultimately the user will understand how many corona hotspots and mainly in which region exists and safely avoid them. This application saves a lot of lives and further helps in decreasing the rate of spread of coronavirus and other similar viruses by alerting them in advance.

KEYWORDS

- **COVID-19**
- **pandemic**
- **hotspot zones**
- **corona zones**
- **crowded areas**

REFERENCES

1. Lovreglio, R.; Ronchi, E.; Kinsey, M. J. An Online Survey of Pedestrian Evacuation Model Usage and Users. *Fire Technol.* **2019**. DOI: 10.1007/s10694-019-00923-8.
2. Yu, I. T.; et al. Evidence of Airborne Transmission of the Severe Acute Respiratory Syndrome Virus. *N. Engl. J. Med.* **2004,** *350* (17), 1731–1739.
3. Lewis, D. Is the Coronavirus Airborne? Experts Can't Agree. *Nature* **2020,** *580* (7802), 175–175. DOI: 10.1038/d41586-020-00974-w.
4. Napoli, M. L. In *Beginning Flutter: A Hands-On Guide to App Development*, 1st Edition, 2020.
5. Shaukat, N.; Ali, D.; Razzak, J. Physical and Mental Health Impacts of COVID-19 on Healthcare Workers: A Scoping Review. *Int. J. Emerg. Med.* **2020,** *13* (1). DOI: 10.1186/s12245-020-00299-5
6. Chaudhary, M.; Sodani, P.; Das, S. Effect of COVID-19 on Economy in India: Some Reflections for Policy and Programme. *J. Health Manag.* 2020, *22* (2), 169–180. DOI: 10.1177/0972063420935541
7. Kharas, H. The Impact of COVID-19 on Global Extreme Poverty [Online]. https://www.brookings.edu/blog/future-development/2020/10/21/the-impact-of-covid-19-on-global-extreme-poverty/(accessed Jan 10, 2021)
8. Burn, W.; Mudholkar, S. Impact of COVID-19 on Mental Health: Update from the United Kingdom. *Indian J. Psychiatry* 2020, *62* (9), 365. DOI: 10.4103/psychiatry.indianjpsychiatry_937_20
9. COVID-19 and its Impact on Education System in India - India Legal. [Online] 2021. https://www.indialegallive.com/legal/covid-19-and-its-impact-on-education-system-in-india/
10. (accessed Jan 10, 2021)
11. Mubarak, N.; Zin, C. Religious Tourism and Mass Religious Gatherings — The Potential Link in the Spread of COVID-19. Current Perspective and Future Implications. *Travel Med. Infect. Dis.* 2020, *36*, 101786. DOI: 10.1016/j.tmaid.2020.101786
12. Kumar, A.; Rajasekharan Nayar, K.; Koya, S. COVID-19: Challenges and its Consequences for Rural Health Care in India. *Public Health Pract.* **2020,** *1*, 100009. DOI: 10.1016/j.puhip.2020.100009
13. Covid-19 Recovery Rate in India Among Highest in the World: Health Ministry-ET HealthWorld. [Online] 2021. https://health.economictimes.indiatimes.com/news/industry/covid-19-recovery-rate-in-india-among-highest-in-the-world-health-ministry/79779406 (accessed Jan 10, 2021)
14. Ghosh, A.; Nundy, S.; Mallick, T. How India is Dealing With COVID-19 Pandemic. *Sens. Int.* 2020, *1*, 100021. DOI: 10.1016/j.sintl.2020.100021
15. Aggarwal, A.; Gaba, S.; Nagpal, S.; Arya, A. A Deep Analysis on the Role of Deep Learning Models Using Generative Adversarial Networks. In *Blockchain and Deep Learning*; Springer: Cham, 2022; pp 179–197.
16. Lovreglio, R.; Ronchi, E.; Kinsey, M. J. An Online Survey of Pedestrian Evacuation Model Usage and Users. *Fire Technol.* 2019. DOI: 10.1007/s10694-019-00923-8.
17. Yu, I. T.; et al. Evidence of Airborne Transmission of the Severe Acute Respiratory Syndrome Virus. *New Engl. J. Med.* 2004, *350* (17), 1731–1739.
18. Lewis, D. Is the Coronavirus Airborne? Experts Can't Agree. *Nature* 2020, *580* (7802),175–175. DOI: 10.1038/d41586-020-00974-w.

19. World Health Organization. COVID-19: Physical Distancing; 2020. https://www.who.int/ westernpacific/emergencies/covid-19/information/physical-distancing.
20. Tian, H.; Liu, Y.; Li, Y.; Wu, C. H.; Chen, B.; Kraemer, M. U. G.; et al. An Investigation of Transmission Control Measures During the First 50 days of the COVID-19 Epidemic in China. *Science* 2020, *368* (6491), 638. https://doi.org/10.1126/science.abb6105 PMID: 32234804
21. Rader, B.; Scarpino, S.; Nande, A.; Hill, A.; Reiner, R.; Pigott, D.; et al. Crowding and the Epidemic Intensity of COVID-19 Transmission. medRxiv. 2020. https://doi.org/10.1101/2020.04.15.20064980
22. Adrian, J.; Bode, N.; Amos, M.; Baratchi, M.; Beermann, M.; Boltes, M.; Corbetta, A.; Dezecache, G.; Drury, J.; Fu, Z.; Geraerts, R.; Gwynne, S.; Hofinger, G.; Hunt, A.; Kanters, T.; Kneidl, A.; Konya, K.; Köster, G.; Küpper, M.; Wijermans, N. A Glossary for Research on Human Crowd Dynamics. *Collective Dynamics* 2019, *4*. https://doi.org/10.17815/CD.2019.19
23. Anderson, R. M.; Heesterbeek, H.; Klinkenberg, D.; Hollingsworth, T. D. How Will Country-Based Mitigation Measures Influence the Course of the COVID-19 Epidemic? *Lancet* 2020, *395* (10228), 931–934. https://doi.org/10.1016/S0140-6736(20)30567-5
24. Bonell, C.; Michie, S.; Reicher, S.; West, R.; Bear, L.; Yardley, L.; Curtis, V.; Amlôt, R.; Rubin, G. J. Harnessing Behavioural Science in Public Health Campaigns to Maintain 'Social Distancing in Response to the COVID-19 Pandemic: Key Principles. *J. Epidemiol. Community Health* 2020, *74* (8), 617–619.
25. Mohler, G.; Bertozzi, A. L.; Carter, J.; Short, M. B.; Sledge, D.; Tita, G. E.; ... Brantingham, P. J. Impact of Social Distancing During COVID-19 Pandemic on Crime in Indianapolis. *J. Crim. Justice* 2020, *68*, 101692.
26. Obradovich Technique For Effective Navigation Based On User Preferences. (2003). Patent No. US 6,542,812 B1, 2003.
27. Huang, J.; Wang, H.; Fan, M.; Zhuo, A. In *Understanding the Impact of the COVID-19 Pandemic on Transportation-related Behaviors With Human Mobility Data*, Proceedings of the 26th ACM SIGKDD International Conference on Knowledge Discovery & Data Mining. https://dl.acm.org/doi/abs/10.1145/3394486.3412856 (accessed Feb 8, 2021)
28. Chhetri, B.; et al. Estimating the Prevalence of Stress Among Indian Students During the COVID-19 Pandemic: A Cross-Sectional Study from India. *J. Taibah Univ. Med. Sci.* 2021, *16* (2), 260–267. https://doi.org/10.1016/j.jtumed.2020.12.012

Tulshyan, V.; Sharma, D.; Mittal, M. An Eye on the Future of COVID'19: Prediction of Likely Positive Cases and Fatality in India over A 30 Days Horizon using Prophet Model. *Disaster Med. Public Health Prep.* 2020, 1–20.

Khosla, P. K.; Mittal, M.; Sharma, D.; Goyal, L. M. Eds.; Predictive and Preventive Measures for Covid-19 Pandemic. In *Algorithms for Intelligent Systems*; Springer, 2021.

Arora, M.; Goyal, L. M.; Chintalapudi, N.; Mittal, M. In *Factors Affecting Digital Education During COVID-19: A Statistical Modeling Approach*, 2020 5th International Conference on Computing, Communication and Security (ICCCS); IEEE, 2020; pp 1–5.

Kaur,A.;Mittal,N.;Khosla,P.K.;Mittal,M.In*MachineLearningToolstoPredicttheImpact of uarantine*; Springer: Singapore, 2021. https://doi.org/10.1007/978-981-33-4236-1_17

Mittal, M.; Battineni, G.; Goyal, L. M.; Chhetri, B.; Oberoi, S. V.; Chintalapudi, N.; Amenta, F. Cloud-Based Framework to Mitigate the Impact of COVID-19 on Seafarers' Mental Health. *Int. Marit. Health* 2020, *71* (3), 213–214.

29. Kanungo, S.; Sharma, D.; Aggarwal, A. COVID-19: Impact and New Normal. *Int. J. Secur. Priv. Perv. Comput. (IJSPPC)* 2021, *13* (1), 57–68.
30. Jain, S.; Aggarwal, A.; Mittal, S. A Study: Chikungunya Using Social Media Analytics in Delhi. *Int. J. Secur. Priv. Perv. Comput. (IJSPPC)* 2021, *13* (1), 46–56.
31. Jain, S.; Aggarwal, A. A Study on Big Data in Health Care Industry. *J. Multidiscip. Eng. Technol. 14*(1).
32. Aggarwal, A.; Dhindsa, K. S.; Suri, P. K. Automation in Healthcare: A Forecast and Outcome–Medical IoT and Big Data in Healthcare. In *Applications of Deep Learning and Big IoT on Personalized Healthcare Services*; IGI Global, 2020; pp 55–69.
33. Aggarwal, A.; Gaba, S.; Nagpal, S.; Arya, A. Recognition of Alphanumeric Patterns Using Backpropagation Algorithm for Design and Implementation with ANN. *J. Secur. Priv. Perv. Comput.* (IJSPPC) 2021, *14* (1), 1–11.
34. Gaba, S.; Aggarwal, A.; Nagpal, S.; Kumar, D.; Singh, P. In *A Forecast of Coronary Heart Disease using Proficient Machine Learning Algorithms*, 2021 Sixth International Conference on Image Information Processing (ICIIP), 2021; pp 517–522. DOI: 10.1109/ICIIP53038.2021.9702640.
35. Nagpal, S.; Aggarwal, A.; Gaba, S. In *Privacy and Security Issues in Vehicular Ad Hoc Networks with Preventive Mechanisms*, Proceedings of International Conference on Intelligent Cyber-Physical Systems. Algorithms for Intelligent Systems; Agarwal, B., Rahman, A., Patnaik, S., Poonia, R.C., Eds.; Springer: Singapore, 2022. https://doi.org/10.1007/978-981-16-7136-4_24.
36. Aggarwal, A.; Gaba, S.; Nagpal, S.; Arya, A. A Comparative Analysis Among Task Scheduling for Grouped and Ungrouped Grid Application, 2021.

Index

S

T

U